MOS Digital ICs

by

George Flynn

HOWARD W. SAMS & CO., INC.
INDIANAPOLIS · KANSAS CITY · NEW YORK

FIRST EDITION

FIRST PRINTING—1975

International Standard Book Number: 0-672-21299-4
Library of Congress Catalog Card Number: 75-28961

Preface

In the words of a popular song, this is the "dawning of the age of Aquarius." To those involved in designing and building the new products of today and tommorrow, a slight revision to the "dawning of the age of technology" might seem more pertinent.

It is easy to forget how short a time we have lived in this technological age, assuming that we have even truly entered it. The automobile is not yet 100 years old; the airplane and radiocommunications are about 70 years old; the transistor and the electronic digital computer have yet to see their 30th birthday. While these inventions and many others have already had fantastic effects on life on earth, we can reasonably assume that much more is yet to happen. Technological evolution has only just begun and, if biological evolution is any guide, has a long way to go.

Medical electronics is in its infancy. Modern transportation systems are inadequate and will have to be either rebuilt or else partially replaced with alternatives such as picture telephones and more powerful data communications systems. The economies of the U.S. and other countries must be made more efficient in supplying the needs of the people and in the conservation and use of natural resources.

Fossil fuels such as coal and oil can no longer be wasted but must be used as effectively as possible, either as sources of energy or as raw materials to make the complex chemicals and plastics that the new technology has developed.

While some people would like to go back to an earlier age, this just won't work. We are going to live in the future whether we like it or not, and either we are going to solve the problems that face us or the problems are going to "solve" us.

The metal-oxide semiconductor (MOS) is an example of an evolving technology that will be used to solve some of the problems. Although MOS is not as innovative today as the transistor was when it first appeared, MOS carries transistor techniques to another level of complexity and is still moving toward even more complexity. The new level of complexity is important in itself, since it allows more difficult problems to be solved without a corresponding increase in the number of pieces of hardware that have to be specified, bought, mounted, wired, tested, etc. The challenge of today is to identify and attack those problems that can be solved with today's level of complexity. Automobile traffic control is receiving much attention. Better control not only provides better use of existing roads but, by eliminating unnecessary stops, saves the energy wasted in accelerating to normal speed again.

The effort in this book has been to present the basics of MOS ICs and to explain how a broad range of these devices operates. The book deals primarily with specific devices that are available off the shelf from many manufacturers and distributors. Not every device can be treated, but a wide cross section is included to provide insight into other ICs that use similar circuits and/or logic. A number of topics common to digital logic circuits have been covered previously in *Transistor-Transistor Logic* (published by Howard W. Sams & Co., Inc.) and are not repeated here.

GEORGE FLYNN

Contents

1

MOS Basics

THE WORLD OF MOS

A few years ago the idea that you could put a complete digital voltmeter (dvm) on one chip of silicon would have been considered pretty far out. Today you can buy an integrated circuit that will do all the logic operations for a 4½-digit dvm. You still have to provide a display system—light-emitting diodes, Nixies, etc., plus associated drivers—and a number of other discretionary circuits, but the sophistication of the basic chip is fantastic.

Today you can also buy a single IC that will perform the four basic mathematical operations and provide answers to 12 places. Sets of several chips perform all the complicated operations of the portable calculator and electronic slide rule. Single-chip microprogrammers and microprocessors are beginning to take over the jobs of small computers. Semiconductor memories now pack 4000 bits on one chip that measures less than 0.2 inch on a side.

The widespread availability of so much sophisticated technology is due primarily to the development of the MOS—Metal Oxide Semiconductor. Most of the complex devices just mentioned are made with MOS transistors.

MOS transistors are similar in operation to bipolar transistors but with some significant differences. A major difference is that the MOS transistor is controlled by voltage, whereas the bipolar transistor is controlled by current. To turn on a bipolar transistor, you supply current to the base; to turn on a MOS transistor, you supply voltage to its control electrode, which is called a gate.

When current is driven into the base-emitter circuit of a bipolar transistor, the resistance of the base-emitter circuit must be overcome by an associated base-to-emitter voltage. That is, the input circuit of a bipolar transistor requires both voltage and current—power. As long as the transistor is held on, it consumes base power. The power gain of the device is the ratio of input power to output power.

In a MOS transistor, however, no direct path exists between the gate input and the source (equivalent to emitter) and drain (equivalent to collector). As a result, the device can be held on indefinitely without using any input current. Actually, of course, since no device is perfect, leakage paths equivalent to 10^{12} to 10^{15} ohms allow some current to flow from gate to source and drain. If the gate is at 10 V, the current will be about 10 V/10^{15} ohms = 10^{-14} ampere, and the input power will be 10^{-14} ampere × 10 V = 10^{-13} watt, or 0.0001 nanowatt. Thus, a MOS transistor can be held on indefinitely with almost no input power consumption. Turning on a MOS transistor, however, requires placing a charge on the gate, and charging the gate does consume power. Not much, but some.

Even the transient power needed to turn on a MOS transistor is less than the steady-state power needed to keep a bipolar transistor on. When a transistor is delivering power into a relatively high current load such as a lamp or relay, the saving of power in the input circuit is not significant. When the load is a large semiconductor memory or another transistor in a long signal-processing chain, the saving in power can become very important.

Another important characteristic of the MOS transistor is its small size. You can put 50 MOS transistors in the same area required for one bipolar transistor. Both the small size and low power consumption of MOS transistors work together to allow very complex circuits in a very small area. The major price paid for the advantages of MOS is speed. MOS ICs are anywhere from 10 to 100, or more, times slower than equivalent bipolar ICs. Even so, MOS ICs operate at low megahertz rates, and this is suitable for many applications. When high speed is necessary or desirable, bipolar circuits can be used.

TYPES OF MOS TRANSISTORS

Bipolar transistors are made in two basic types: npn and pnp. Transistor action in both types is similar, and the major difference as far as the external circuit is concerned is that the npn requires a positive voltage on its collector while the pnp requires a negative voltage on its collector. In a MOS transistor we have a similar situation but with two additional complications.

First, the control electrode, or gate, can be formed in one of two ways: either as a pn junction or as an insulated electrode. Second, the transistor can be either a depletion-mode device or an enhancement-mode device. Depletion-mode transistors conduct when the voltage on the control electrode is 0; enhancement-mode transistors do not conduct when the control electrode voltage is 0. As a result we have six basic MOS transistors.

Consider first the bipolar npn transistor shown in Fig. 1-1A. When base voltage is zero, the two diodes prevent current flow between collector and emitter. As base voltage is increased from

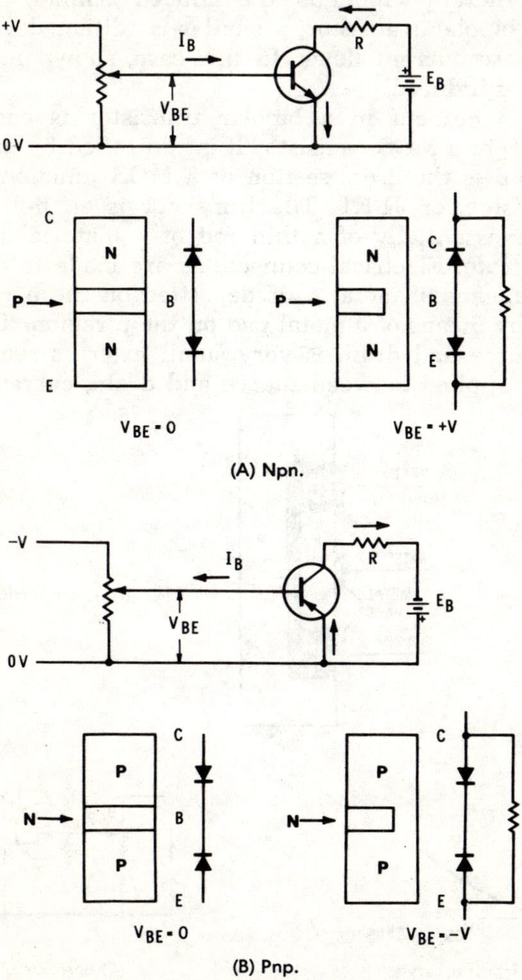

(A) Npn.

(B) Pnp.

Fig. 1-1. Bipolar transistors.

0 toward +V, base current begins to flow and the transistor begins to turn on, allowing current to flow within the transistor from collector to emitter. What has happened, in effect, is that the flow of base current has opened up a conducting path, or channel, from collector to emitter. The more base current there is, the wider the conducting channel becomes and the lower its resistance. Thus, in the conducting state the npn transistor looks like two conducting n regions connected together by a conducting channel. Current cannot flow from emitter to collector, however, without at the same time reducing base-current flow to the emitter and thereby wiping out the induced channel.

In a pnp bipolar transistor, a similar condition develops when emitter-to-base current flows. In this case, shown in Fig. 1-1B, a p channel is induced.

While load current in a bipolar transistor is controlled by base current, in a MOS transistor it is controlled by gate voltage. Fig. 1-2A shows the cross section of a MOS junction-type field-effect transistor, or JFET. The transistor is an n-channel type and consists essentially of a thin rod of n material buried in a p-type substrate. Electrical connections are made to each end of the rod by means of metal pads deposited on the n regions, and to the gate by means of a metal pad on the p region. The channel between source and drain is very small in cross section. When a voltage is applied between source and drain, current will flow

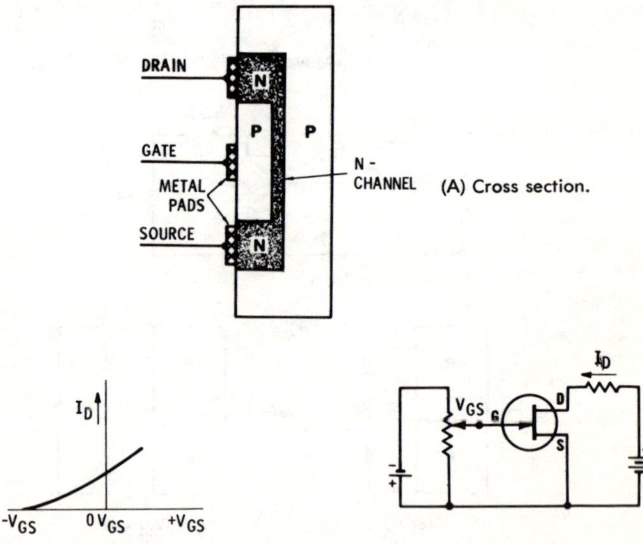

(A) Cross section.

(B) Transfer curve.

(C) Basic circuit.

Fig. 1-2. N-channel JFET.

through the device. Since the current flows only through n-type material and not through any pn junction, the current will flow in either direction, depending on the polarity of the applied voltage. Thus, either end of the rod can act as source or drain.

The transfer curve for an n-channel JFET is shown in Fig. 1-2B, and a basic circuit at C. When the gate-to-source voltage is 0, load current flows. If V_{GS} is made positive, the pn junction is forward-biased and the device acts like a diode. Thus, in normal operation the gate voltage is negative. In this region the pn junction is reverse-biased and the input impedance is very high.

Because of the voltages applied to the device, a strong electrical field exists at the boundary of the pn junction. The field is strongest near the drain contact and weakest near the source. Since the p material is negatively biased, it attempts to drive electrons into the n channel; but the n channel already has an excess of electrons that are repelled by the electrons in the p material. As a result, the n channel becomes squeezed in upon itself and becomes smaller in cross section. The resistance of the channel is increased, and load current is decreased. Thus, the gate-to-source voltage acts to control load current. When the gate is made sufficiently negative, the channel is essentially pinched off and load current is reduced to 0.

A p-channel JFET can be formed as shown in Fig. 1-3A, where a p channel is buried in an n substrate. The pn junction

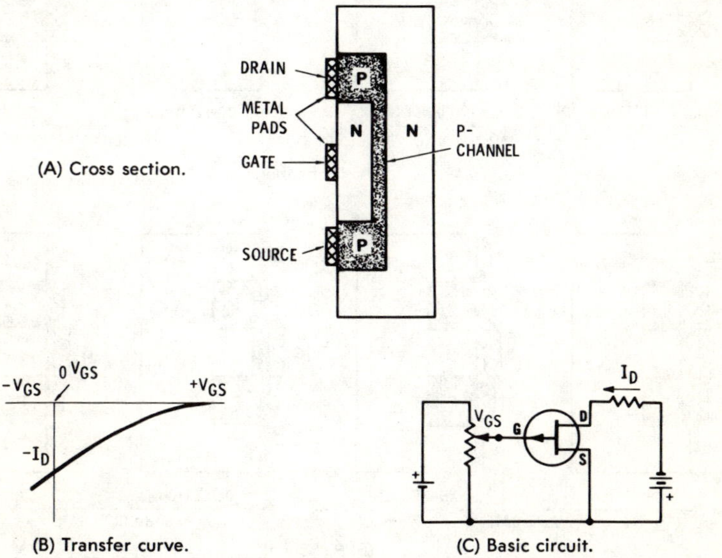

(A) Cross section.

(B) Transfer curve.

(C) Basic circuit.

Fig. 1-3. P-channel JFET.

must not be forward-biased; therefore, the n-material gate region is kept positive with respect to the p-region source and drain. Drain current I_D is shown flowing into the terminal, but the actual current flow is out of the drain. The transfer curve, shown in Fig. 1-3B, therefore plots V_{GS} versus $-I_D$. As V_{GS} is made more and more positive, the electric field acts to choke off the p channel and eventually I_D is reduced to 0.

JFETs were the first MOS devices to be widely applied. The next step was the insulation-gate FET, or IGFET. The cross section of an n-channel IGFET is shown in Fig. 1-4A. We still have n-type material used for source, drain, and channel, with a p-type substrate. Instead of a pn junction at the input, however, we have a very thin layer of insulation—silicon dioxide—between the metal gate and the n channel, and the channel may be formed on the surface of the substrate rather than being buried.

The device operates similarly to the JFET except that now the gate can be made positive as well as negative, since the insulation prevents current flow from gate to source or drain. Breakdown

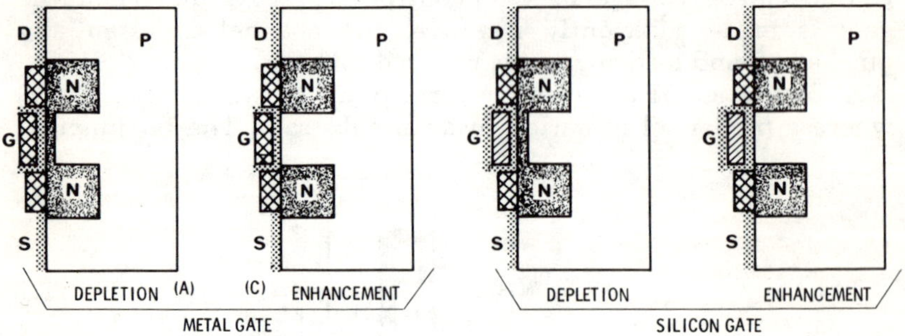

(A) N-channel.

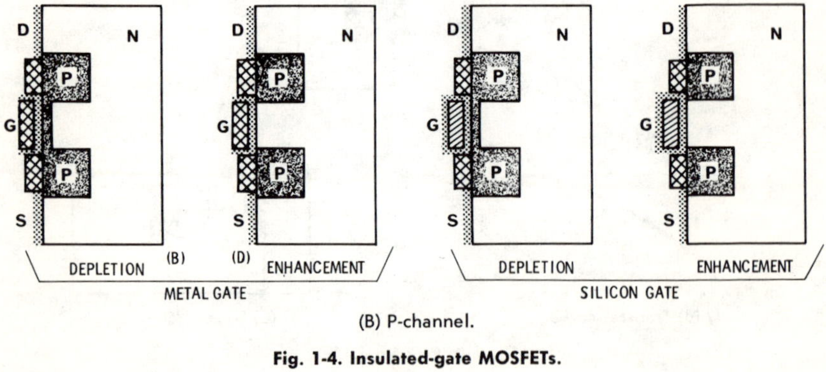

(B) P-channel.

Fig. 1-4. Insulated-gate MOSFETs.

voltage of the gate insulation is usually 25 to 100 V or so, and thus the gate can be directly connected to the same voltages as the sources and drains. The p material of the substrate must be kept negative with respect to the n material. Usually the source is connected to the substrate.

Again, a similar device using p material for source and drain is possible and takes the form shown in Fig. 1-4B. Operation is similar to that of the n-channel just described, except that the polarity of the operating voltage is reversed.

The MOSFETs mentioned so far are all built with a permanent channel and operate in what is called the depletion mode. The next step is a device without a built-in channel but one in which a channel can be induced. Fig. 1-4C shows the cross section of a metal-gate device with n-type source and drain, between which a channel can be created by applying a gate-to-source voltage. The device is called an enhancement-mode n-channel MOSFET. As the gate voltage is made positive, electrons are attracted to the metal gate and holes are driven deeper into the p material. As the gate-to-source voltage is increased from 0, an n-type channel begins to form, starting at the source. As the voltage increases, the channel lengthens until at some critical threshold voltage it extends all the way from source to drain. Once the threshold voltage is reached, conduction from source to drain through the channel begins; load current flows through a purely ohmic or resistive device, not through a pn junction. If the voltage is increased too much, it will punch through the insulation of the gate and the device will be destroyed. Below this value, however, the device acts like a transistor and has the transfer curves shown in Fig. 1-5A.

An equivalent p-channel device can be made, and it takes the form shown in Fig. 1-4D. Transfer curves are shown in Fig. 1-5B.

A major characteristic of the enhancement-mode device is that it draws no current when the input signal is 0. Thus, the operating

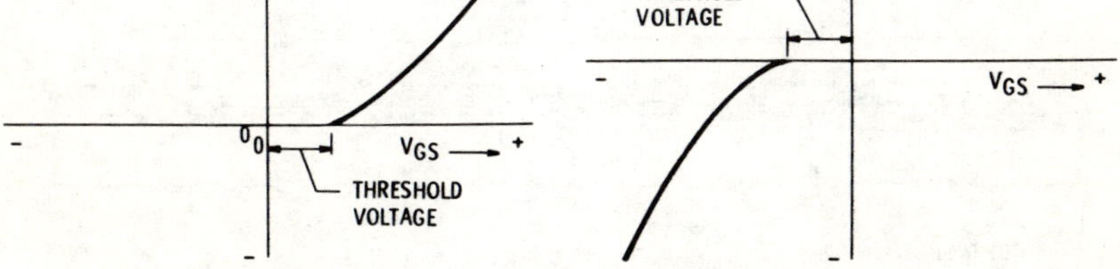

(A) N-channel.

(B) P-channel.

Fig. 1-5. Transfer curves of enhancement-mode MOS transistors.

power is close to 0 when the device is quiescent, or turned off. This is a major advantage in digital circuits, where about half the time the input signal will be a 0.

For the metal-gate enhancement-mode transistor to work, the metal gate must be accurately placed between source and drain and must also overlap source and drain slightly so that the induced channel will extend all the way from source to drain. Any extra overlap, however, adds capacitance to the device and slows its speed of response. To get around this problem, the silicon gate structure was developed. The silicon gate is made from a special material called polysilicon and is formed at an early stage in the manufacturing process. Then the source and drain are formed by diffusion, and at the same time dopant atoms are diffused into the polysilicon gate material to make it conductive. Since the gate is already in place when source and drain are formed, the gate is said to be self-aligning. Gate, source, and drain are all perfectly aligned, with no overlap and therefore no excess capacitance. Since no allowance need be made for gate misalignment, the device can be made still smaller, again decreasing capacitance and increasing speed. Cross sections, approximately to scale, of metal and silicon-gate transistors are shown in Fig. 1-6.

The silicon-gate structure is the most widely used MOS transistor in digital ICs today and is being used in more and more circuits. Enhancement-mode devices give the lowest power dissipation and thus the least heat generation and the best promise of reliability. Metal-gate transistors are also being made in large quantities, since the manufacturing processes are well under control and a lot of money has been invested in masks and production equipment. Junction-type MOSFETs are not used in digital ICs because they are depletion-mode devices and would dissipate power during quiescent operation; they are used in special cases where their particular characteristics are wanted.

Many symbols have been created to represent MOS transistors, and Fig. 1-7 is a summary of them. The symbol shown at G is the most widely used, since it has enough detail to be distinctive yet is simple and easy to draw.

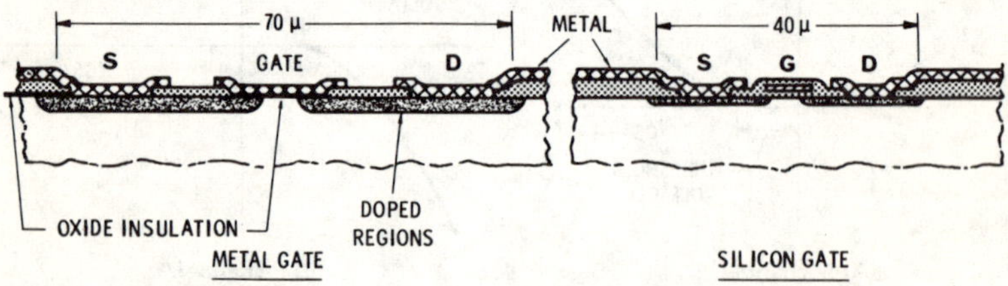

Fig. 1-6. Cross-section view of metal-gate and silicon-gate transistors.

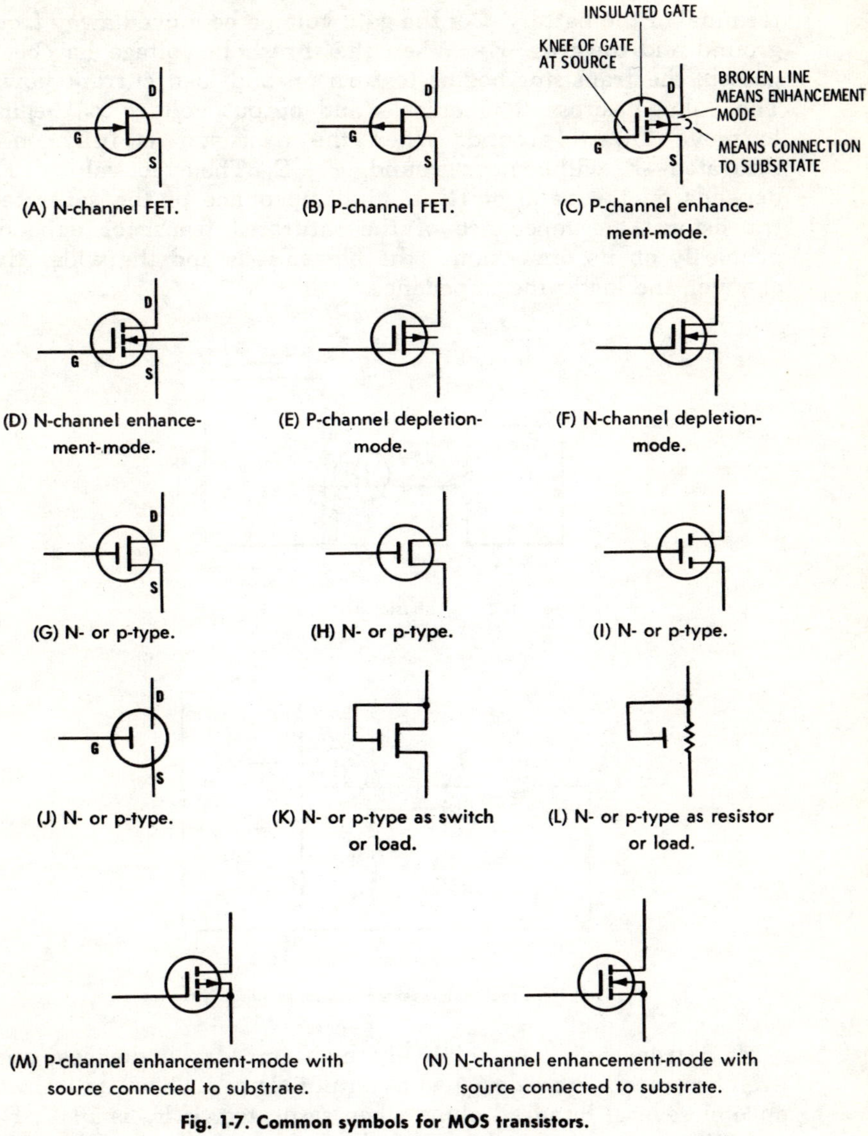

(A) N-channel FET.

(B) P-channel FET.

(C) P-channel enhancement-mode.

(D) N-channel enhancement-mode.

(E) P-channel depletion-mode.

(F) N-channel depletion-mode.

(G) N- or p-type.

(H) N- or p-type.

(I) N- or p-type.

(J) N- or p-type.

(K) N- or p-type as switch or load.

(L) N- or p-type as resistor or load.

(M) P-channel enhancement-mode with source connected to substrate.

(N) N-channel enhancement-mode with source connected to substrate.

Fig. 1-7. Common symbols for MOS transistors.

MOS AMPLIFIER-INVERTER

The basic circuit for a p-channel enhancement-mode MOS transistor is shown in Fig. 1-8A. When the gate is at the same potential as the source, the device is cut off, no load current flows, and output voltage E_O is at the same potential as the negative terminal of the battery. Let the gate voltage be moved away from ground and toward $-E_X$. When the threshold voltage has been passed, the transistor begins to turn on and load current flows. The voltage across R_L increases and output voltage E_O begins to move toward ground. When the transistor is fully on—saturated—E_O will be near ground, or $+E_B$. The exact value of E_O depends on the ratio of R_L to the impedance of the saturated transistor. The impedance of the saturated transistor depends primarily on its dimensions: the bigger it is and the wider the channel, the lower the impedance.

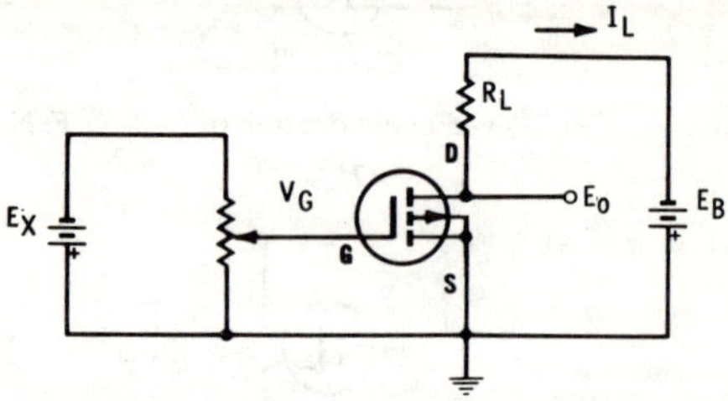

(A) Basic circuit.

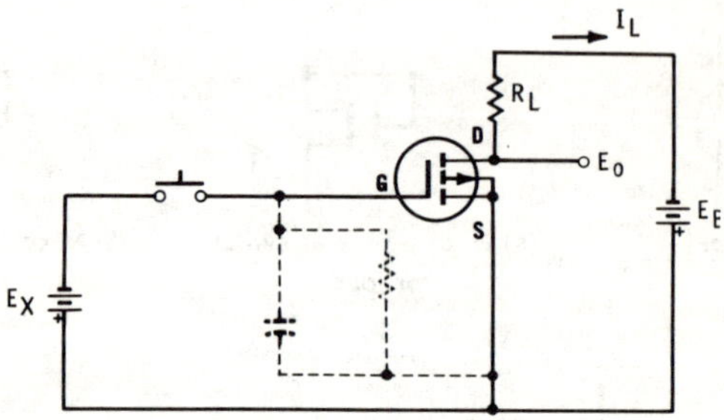

(B) Gate capacitance and leakage.

Fig. 1-8. P-channel, enhancement-mode, amplifier/inverter.

MOS transistors are relatively high impedance devices, and even large transistors designed as output elements have an impedance of several hundred ohms when saturated. If E_B is 10 V, R_L is 100K, and the impedance of the saturated transistor is 500 ohms; E_O can swing from −10 V when the transistor is off to −0.0497 when it is on.

The voltage gain of the circuit is high, but not as high as it would be for an equivalent bipolar transistor. Even so, only a small change in V_G is needed to swing E_O from $-E_B$ to near ground. The power gain is very large.

Consider the circuit of Fig. 1-8B. When the push button is closed, the transistor turns on. Note, also, that associated with the gate is a small gate-to-source capacitance which becomes charged when the push button is closed. When the push button is released, the charge on the gate begins to leak off through the gate-to-source insulation. While this charge is leaking off, however, the transistor remains on. If the push button is closed again before too much charge leaks off, the transistor will remain on. The gate-to-source capacitance is only a few picofarads, but the leakage resistance is very high. For a typical example, an IC might have an input capacitance of 5 pF and a leakage resistance of 10^{12} ohms; then the RC time constant would be 5 seconds. The full 5 seconds cannot be used if the transistor is to be kept on continuously, but a substantial amount of time, when measured in terms of digital clock rates, can be allowed to pass without recharging—or refreshing—the gate capacitance. This technique is used in dynamic shift registers and memories, which will be discussed later.

The basic circuit provides amplification and voltage inversion. The circuit can be direct-connected as shown in Fig. 1-9A, to add another inversion. In ICs, amplifier chains such as the one in Fig. 1-9A can be realized easily. Transistors Q1 and Q2 can be made very small, but resistors R1 and R2 use up a relatively large amount of silicon real estate. Another MOS transistor, however, permanently biased somewhere into its conduction region, will look like a large resistor to Q1. The circuit takes the form shown in Fig. 1-9B.

Transistors Q_{L1} and Q_{L2} function as loads for Q1 and Q2, respectively, and are often shown as in the NMOS inverter circuit of Fig. 1-9C.

The gate of the load transistor is permanently tied to the drain, which turns the transistor on. The impedance of Q_{L1} must be much larger than the "on" impedance of Q1 to ensure that point X has an adequate voltage swing. In terms of silicon real estate, Q1 is considerably larger than Q_{L1}.

When the source of Q_{L1} is riding near $-E_B$ (Fig. 1-9B), Q_{L1} has a certain impedance; when the source is near $+E_B$, it has another impedance. That is, the resistance of Q_{L1} is nonlinear. This nonlinearity is acceptable in digital circuits as long as a signal can be propagated down the chain. This condition is easily met, and most MOS digital ICs consist of nothing but transistors.

Inverters can also be made by using one n- and one p-channel transistor in a complementary structure, as shown in Fig. 1-9D. Note in this circuit that the source of the p-channel transistor is connected to the plus terminal of E_B, while the source of the

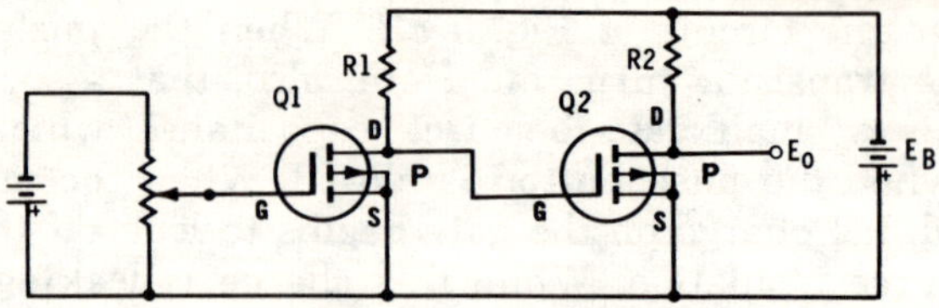

(A) Two-stage amplifier.

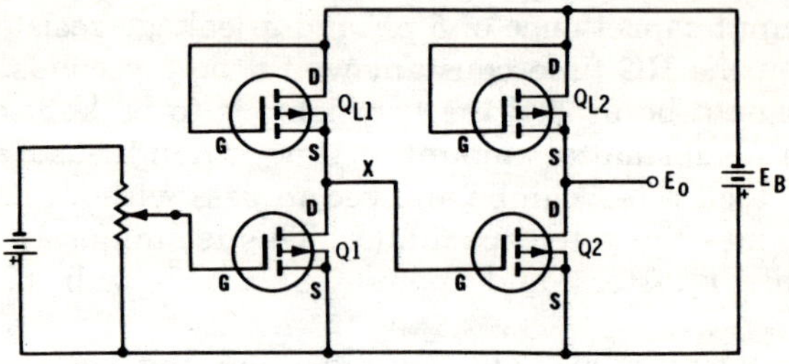

(B) MOS transistor used for load resistor.

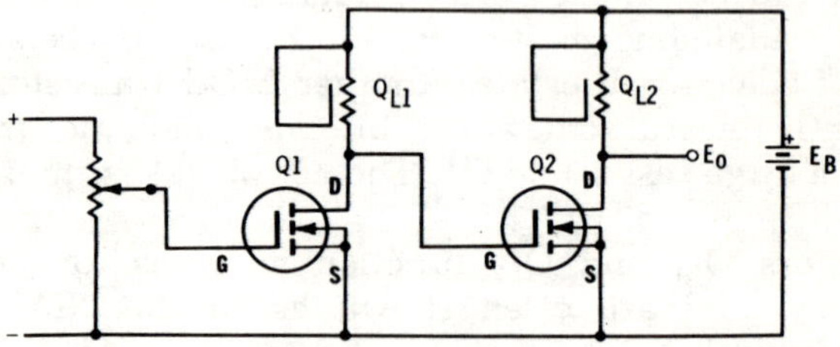

(C) NMOS inverters.

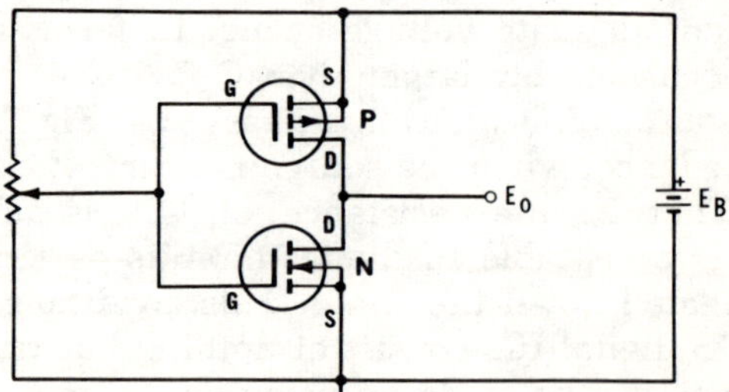

(D) Complementary inverter.

Fig. 1-9. MOS amplifiers.

n-channel device is connected to the minus terminal, with the drains connected together and tied to the output.

When the gate is connected to a positive voltage, the n-channel transistor has a high gate-to-source potential and is therefore turned on, while the p-channel transistor has its source and gate at the same potential and is therefore turned off. Output voltage E_O with QN on is close to 0 V.

When the gate voltage is switched to ground, the reverse action occurs. The n-channel transistor now has gate and source at the same potential and is turned off, while the p-channel transistor has a high source-to-gate potential and is turned on. Output E_O goes high, close to $+E_B$.

What happens when the gate input is not connected to $+E_B$ or $-E_B$ but is allowed to float? Electrons will be drawn from the gates through the insulation of the p-channel transistor toward the plus terminal of the supply voltage, while others will be fed to the gate through the insulation of the n-channel transistor. If the gate becomes positively charged, the n-channel transistor turns on; if electrons accumulate on the gates, the p-channel transistor turns on. The output is unpredictable. The gates must be connected to +V or −V for reliable operation.

MOS FAMILIES

Because of the relatively large number of basic types of MOS transistors, a relatively large number of MOS digital logic families have been created. The most important of these are based on three circuit approaches: p-channel enhancement-mode circuits, n-channel enhancement-mode circuits, and complementary circuits using p- and n-channel enhancement-mode devices.

The first MOS ICs to become widely available used p-channel enhancement-mode transistors and a relatively high negative voltage (−28, −40, etc.), and most require more than one voltage supply for proper operation. Many require quite a lot of external electronics. Memory-type circuits are favored and include dynamic shift registers and dynamic memories, some of which require 2, 3, or even 4 clock signals.

The next group of MOS ICs to be developed was based on the complementary circuit, CMOS, and three major families of this class have so far arrived in force. These are the RCA 4000 series, National Semiconductor 54C/74C, and Motorola 14500. (A 54CXX device is rated to operate from −55 to +125°C, while a 74CXX is rated from −40 to +85°C. The narrower span is called a commercial rating and the wider, a military rating. Similar distinctions are made in other families using similar codes.)

Devices in all these families are now available from more than one manufacturer. All three families can operate with power supplies of +3 to +15 V (to +18 V in a few cases), and thus they can work with the same +5-V power supplies as TTL circuits. In addition, the 54C/74C duplicates on a pin-for-pin basis many—but not all—devices in the 54/74 families of TTL, and it uses the same package numbers.

A 54C00/74C00 is a quad 2-input NAND gate (CMOS), like the 5400/7400 (regular TTL), 54L00/75L00 (low-power TTL), and 54S00/74S00 (Schottky TTL). A printed-circuit board originally designed for 54/74 can be used to build the same circuit in 54C/74C, using the same IC part numbers—if available. Series 4000 and 14500 devices can also be used to build the same or a similar circuit, but the part numbers will be different and the board will be different. Type 4011, for example, is also a quad 2-input NAND gate, but the input and output pins are not numbered the same as the 54/74 pins.

The 54C/74C family is under a significant restraint in that it is required by definition (or at least by mission) to duplicate 54/74 circuits. The other CMOS families do not have to meet these conditions and thus are free to develop along lines that are inherent in MOS circuits. As a result, numerous circuits that are not possible or practical in TTL have been made available in 4000, 14500, and other CMOS families. Whether or not 54C/74C will rebel and go in similar directions remains to be seen.

The NMOS family, using n-channel transistors, was slow to develop until CMOS ICs proved that the technology had been mastered. NMOS devices have several advantages: they can operate at higher speed than PMOS devices, can work with a single power supply from as low as +1.5 to as high as +15 V or so, and can provide even greater circuit density.

MOS technology is still advancing, and significant improvement in performance, complexity, and ease of use is certain. Among the most important possibilities are ion implantation, silicon on sapphire, and D-MOS or double-diffused MOS. Ion implantation provides much better control of device characteristics than thermal diffusion and is already in wide use. Silicon on sapphire—SOS—allows very thin source and drain regions, which reduces capacitance and increases speed, but the technology is in its early phases and has been beset by problems. D-MOS has been used to make individual transistors that work at frequencies in the gigahertz region, but whether it will be used in MOS ICs remains to be seen.

2

CMOS

NAND GATES IN CMOS

NAND gates perform the logic function of AND, plus inversion. The NAND function can be indicated by any of the ways shown in Fig. 2-1. The D symbol at Fig. 2-1A represents the AND function, and the open circle or bubble is used to show negation or inversion. The symbol at B is the same as at A except that the open triangle replaces the open circle. The open triangle is meant to be interpreted as a polarity indicator.

In a straight AND gate, the output is true if inputs A and B are true. The symbol at B implies that if A and B are true, Y is true but low. In effect, then, passing through an open triangle means that the logic scheme has changed from positive logic (high voltage = 1, or true; low voltage = 0, or not true) to negative logic, where low signals are true and high signals are not true. Passing through another open triangle at the input of another circuit means that a low input is true. So far as a logic diagram is concerned, both symbols represent exactly the same thing. The same information can be represented by the Boolean equation in Fig. 2-1C and the truth table in Fig. 2-2.

A 2-input NAND gate in CMOS is shown in Fig. 2-3. P-channel transistors are on top with their sources connected to V_{CC} (which is also called V_{DD} by some manufacturers) and their drains connected to the output. The n-channel transistors are on the bottom in series, with the top drain connected to the output and the bottom source to ground or V_{SS}, which is the substrate. (Voltages that are fed to multiple devices or ICs are usually identified with a double letter.)

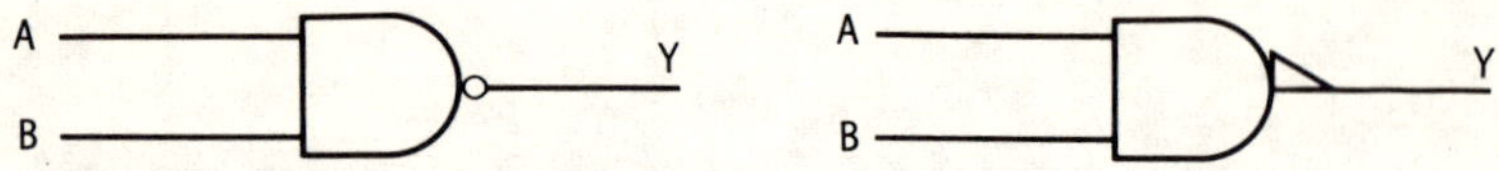

(A) Symbol showing negation or inversion. (B) Symbol with polarity indicator.

$\overline{Y} = A \cdot B$ (C) Boolean equation.

Fig. 2-1. CMOS NAND gate.

A	B	Y
0	0	1
0	1	1
1	0	1
1	1	0

A	B	Y
L	L	H
L	H	H
H	L	H
H	H	L

A	B	Y
X	0	1
0	X	1
1	1	0

1 = LOGIC 1
0 = LOGIC 0
L = LOW
H = HIGH
X = IRRELEVANT

Fig. 2-2. Truth tables for CMOS NAND gate.

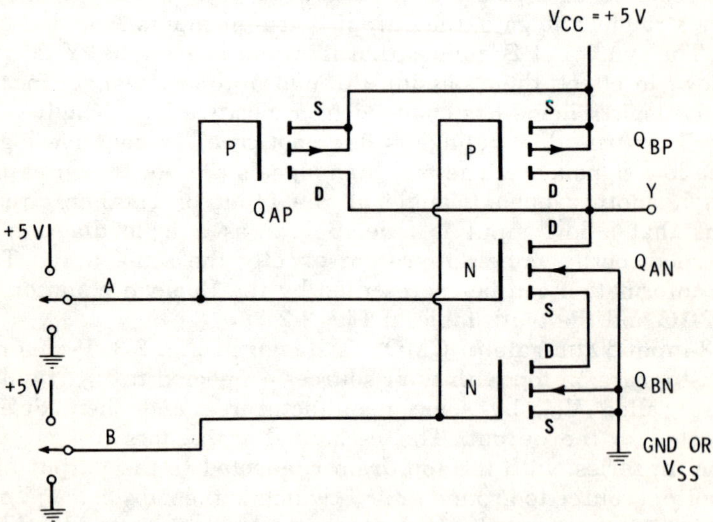

Fig. 2-3. Two-input CMOS NAND gate (54C00/74C00:4011).

Let inputs A and B both be connected to ground. The bottom n transistor has gate and source at the same voltage and therefore is off. The upper n transistor has its gate at ground potential, which is even lower than its source; thus, it also will be off.

Both p transistors, on the other hand, have the correct potential between gate and source to turn on. As a result, output Y is high. If no load current is drawn, output Y is within a millivolt of V_{CC} since the only impedance to the flow of current from V_{CC} to Y is the purely ohmic resistance of two turned-on p-channel transistors in parallel. The drop across these transistors is in direct proportion to load current, and if the load current is zero, the drop is zero.

Now let input A be set to +5 V (= V_{CC}) while B remains at 0 V. Transistor Q_{AN} now has the correct gate voltage to turn on, but its source is not connected to ground; transistor Q_{AP} now has the wrong voltage for operation and is turned off. Since Q_{BP} is still turned on by input B, output Y stays high. Change the inputs so that A is low and B is high. Now, Q_{AP} and Q_{BN} are on but Q_{BP} and Q_{AN} are off. Output Y remains high.

Set both inputs A and B high. Transistor Q_{BN} is on; Q_{AN} has its source connected to ground through Q_{BN} and is on; and Q_{AP} and Q_{BP} both have the wrong gate voltage and are off. Thus, output Y goes low.

Output Y can serve as a voltage source to a high-impedance load, since it will be at about 4.99 V when high. When Y is low, it delivers 0 V or can act as a current sink to another load such as a TTL gate.

Several peculiarities of the CMOS NAND gate should be noted. When Y is high, either Q_{AP} is on, or Q_{BP} is on, or both are on. When Q_{AP} is on, it has a certain impedance; Q_{BP} has a close but slightly different impedance when on, and when both are on, their combined parallel impedance is only about one-half that of either alone. If Y is feeding a high-impedance load, this change in source impedance may not be of any significance, but if the load impedance is not high, the change may need to be considered.

A second point to note is that when Y is low, it is connected to ground through two transistors in series. If Y is acting as a voltage source, no problems result. If Y is acting as a current sink, load current will flow through both transistors and the resulting voltage drop will cause the output voltage to rise. In addition, the voltage at the source of Q_{AN} will rise, and this means that input A must be high enough to overcome this extra voltage. This effect is not serious in a 2-input NAND gate. As the number of inputs increases, however, the circuit becomes more sensitive to load current. Fig. 2-4 shows a 4-input NAND gate; when Y is low, load current must flow through four transistors in series to ground. When Y is high, the output impedance may consist of just one turned-on transistor, or 2, 3, or 4 in parallel.

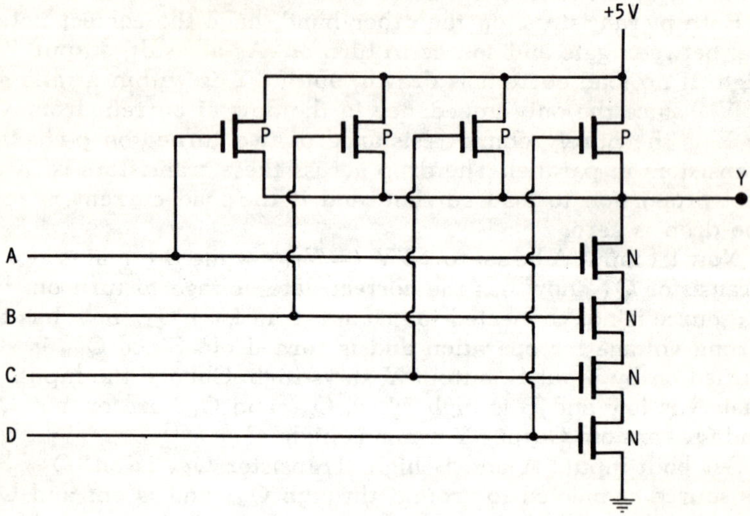

(A) Circuit.

A	B	C	D	Y
X	X	X	0	1
X	X	0	X	1
X	0	X	X	1
0	X	X	X	1
1	1	1	1	0

X = IRRELEVANT

$$\overline{Y} = A \cdot B \cdot C \cdot D$$

(B) Truth table and Boolean equation.

Fig. 2-4. Four-input CMOS NAND gate.

Because of these effects, CMOS NAND gates are not made with more than 4 or 5 inputs. While type 54C/74C30 is an 8-input NAND, it uses two levels of gates, as shown in Fig. 2-5 (shown with all inputs = 1; if any input goes to 0, Y goes to 1). When a large number of inputs are needed, gates can be combined in similar ways.

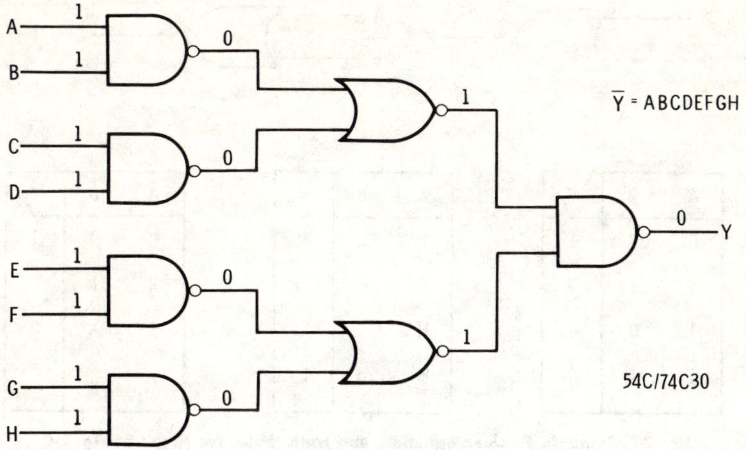

Fig. 2-5. Eight-input CMOS NAND gate.

NOR GATES IN CMOS

The circuit for a 2-input NOR gate in CMOS is shown in Fig. 2-6 and is very similar to the NAND gate circuit. For NOR, the n-channel transistors are connected in parallel while the p-channel transistors are connected in series; for NAND, the n-channels are in series and the p-channels are in parallel. The symbols and truth tables are shown in Fig. 2-7.

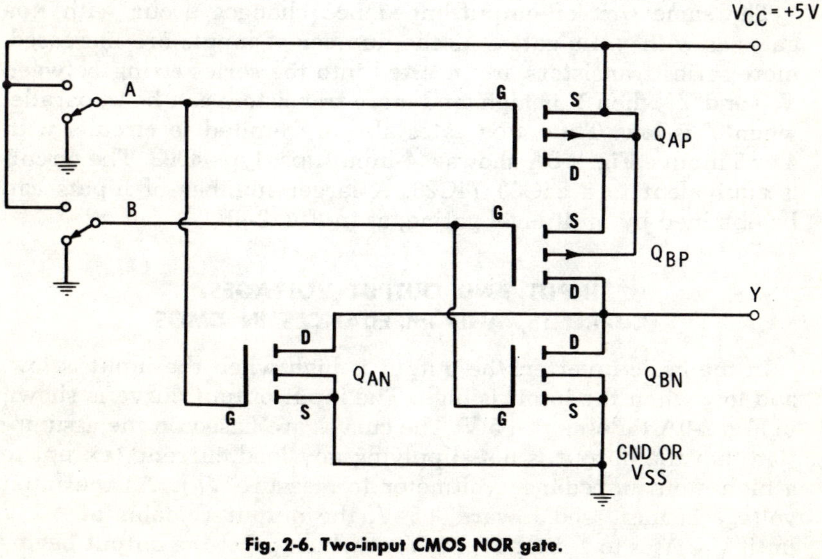

Fig. 2-6. Two-input CMOS NOR gate.

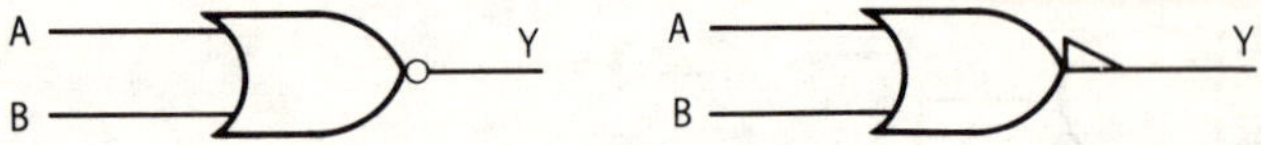

$\overline{Y} = A + B$

A	B	Y
0	0	1
0	1	0
1	0	0
1	1	0

A	B	Y
L	L	H
L	H	L
H	L	L
H	H	L

A	B	Y
0	0	1
X	1	0
1	X	0

Fig. 2-7. Symbols, Boolean equation, and truth tables for circuit of Fig. 2-6.

When A and B are both 0, p-channel transistors Q_{AP} and Q_{BP} both have the correct gate-to-source voltage needed to turn on, and both n-channel transistors Q_{AN} and Q_{BN} are turned off. Thus, output Y is high. If input A is connected to $+V_{CC}$, Q_{AP} turns off and Q_{AN} turns on. With Q_{AP} off, Y is disconnected from V_{CC}, and since Q_{AN} is on, Y is connected to ground. If input B instead of A is connected to $+V_{CC}$, Q_{BP} opens and Q_{BN} closes, again connecting Y to ground. If both inputs are connected to $+V_{CC}$, both p-channel transistors are open and both n-channel transistors are closed, producing the same result—Y connected to ground.

The same sort of output-impedance changes occur with NOR gates as with NAND gates. As the number of inputs are increased, more series transistors are inserted into the series string between V_{CC} and Y, when Y is high, and more transistors can be in parallel when Y is low. Thus, NOR gates also are limited to circuits with 4 or 5 inputs. Fig. 2-8A shows a 4-input NOR, type 4002. The circuit is equivalent to a 54C20/74C20. A larger number of inputs can be obtained by multilevel gating, as in Fig. 2-8B.

INPUT AND OUTPUT VOLTAGES, CURRENTS, AND IMPEDANCES IN CMOS

In the basic inverter, the output is high when the input is low, and low when the input is high. The input-output curve is shown in Fig. 2-9A for $V_{DD} = +5$ V. The curves are based on the assumption that the output is not supplying any load current (except to a high-input-impedance voltmeter to measure V_O). As the input voltage is increased toward +5 V, the output remains at +5 V until V_{IN} rises to a little over 2 V, at which point the output begins

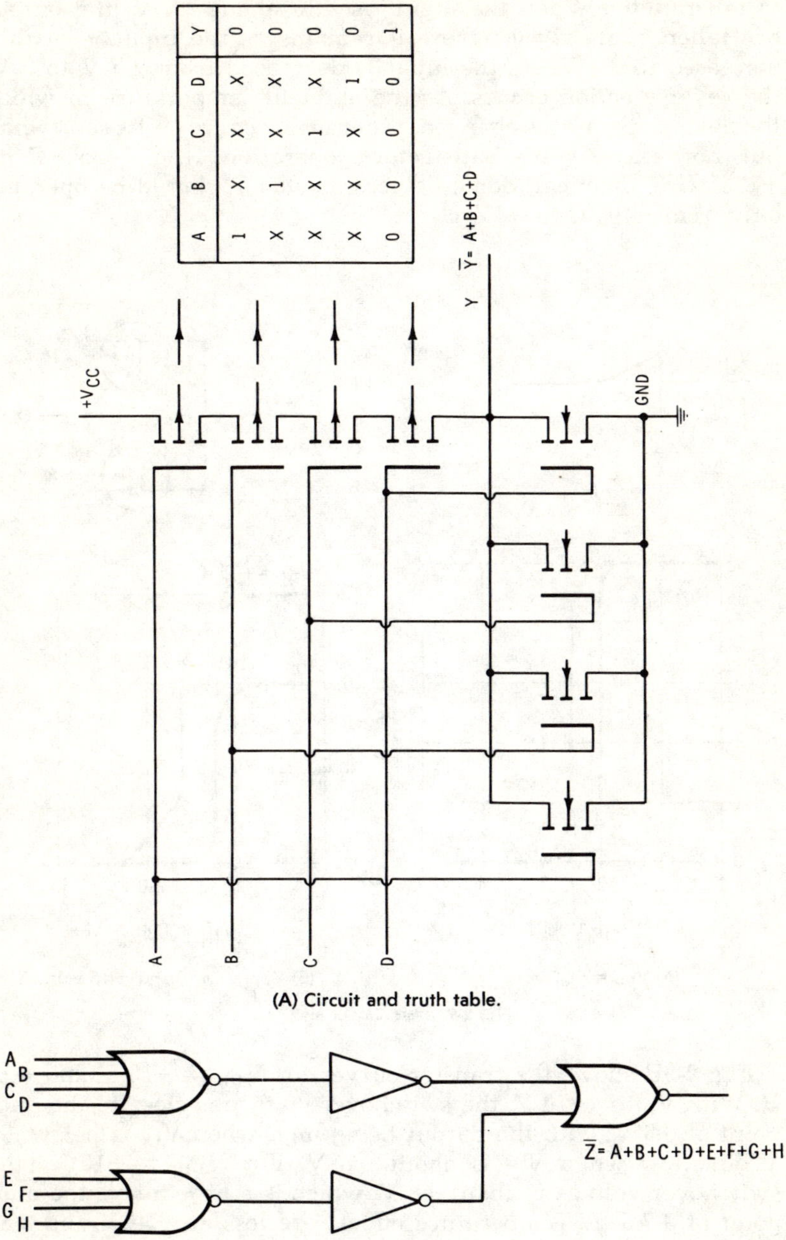

A	B	C	D	Y
1	X	X	X	0
X	1	X	X	0
X	X	1	X	0
X	X	X	1	0
0	0	0	0	1

(A) Circuit and truth table.

(B) Method for obtaining more inputs.

Fig. 2-8. Four-input CMOS NOR gate.

to fall rapidly. When the input rises to about 2.6 V, the output has fallen to nearly 0, where it remains as the input is further increased to +5 V. As the input is decreased from +5 V to 0 V, the reverse action occurs. As the ambient temperature in which the device operates is changed, the transfer curve shifts somewhat, but not excessively. Satisfactory operation from −55° C to +125° C is practical. Some devices, however, should be operated only from −40° C to +85° C.

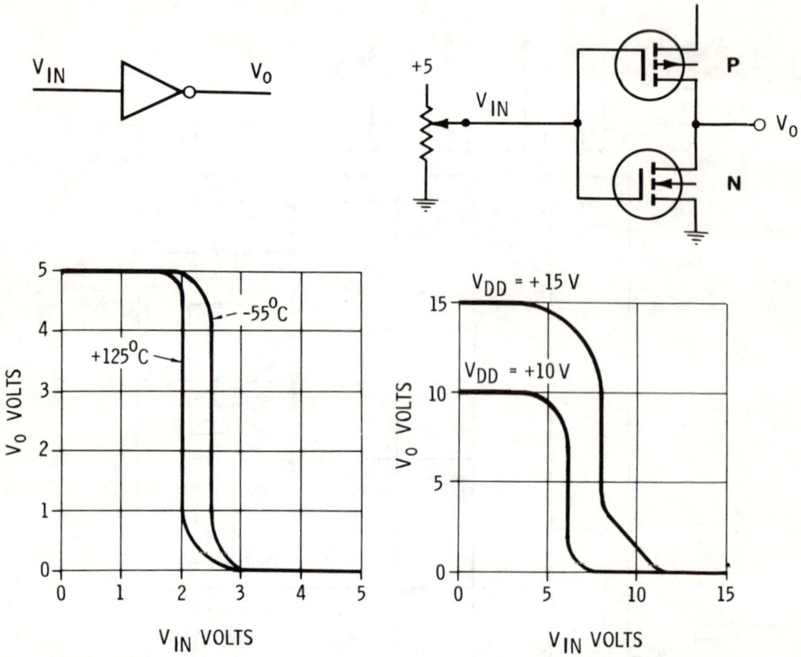

(A) V_{DD} = +5 volts.

(B) V_{DD} = +10 and +15 volts.

Fig. 2-9. Basic CMOS inverter.

Fig. 2-9B shows the transfer curves for V_{DD} = +10 V and +15 V. With V_{DD} = +10 V, the switchover voltage is close to the midpoint of +5 V, with the output being high when V_{IN} is below +5 V, and low when V_{IN} is about +6 V. For V_{DD} = +15 V, the switchover voltage is about +8 V, which is a little above the midpoint of +7.5 V. Temperature effects are less at +10 V and +15 than at +5 V.

Let one CMOS inverter drive another as in Fig. 2-10A. Currents in CMOS circuits are defined as positive when flowing into the device, as shown by the arrows in Figs. 2-10A and B. Thus, an output current, I_O, flowing into G2 is shown as $+I_O$. When V2 is high, current actually flows out of the p-channel transistor of G2 and into R_L. This situation is referred to on data sheets as $-I_O$. Thus, the minus sign is used to reverse the direction of the arrow.

Let V_{IN} be below 2 V. Then V1 will be high, G1 will be driving G2, and V2 will be low. What will be the actual current and voltages existing in the circuit? Gate G1 must drive the input of G2. The input of G2 can be represented as shown in Fig. 2-10C. Each inverter has an input capacitance of about 5 pF in parallel with a resistance of about 10^{12} ohms that represents leakage current. Once the capacitance has been charged, no further current flows into it. If V1 is 5 V, the current through R_N is $I = 5\ V/10^{12}$ ohms $= 5 \times 10^{-12}$ ampere = 5 femtoamperes = 0.005 nanoampere. In other words, once the input capacitance of G2 has been charged, no further input current is needed. Since transistor P1 of G1 is on but does not have to supply any current to G2, output voltage V1 is essentially the same as V_{DD}, or +5 V.

Voltage V1 acts to turn on the n-channel transistor of G2 and thus V2 is brought to ground. Since R_L is connected to ground, V2 is 0. The input current to G1 is determined by V_{IN}, which is less than +2 V. Thus, the input current to G1 is even less than that to G2.

The only other current flow in the circuit is the leakage current from V_{DD} to V_{SS} through the turned-off N1 and the turned-off P2. The quiescent supply current, as it is called (and usually identified as $+I_{DD}$ or $-I_{SS}$), is a low of about 0.05 μA to a high of about 15 μA, depending on the individual device. Thus, the quiescent power dissipation of the circuit is a minimum of $5\ V \times 0.05 \times 10^{-6} = 25 \times 10^{-8} = 0.25\ \mu W$ and a maximum of $5\ V \times 15 \times 10^{-6} = 75\ \mu W$. This range of 75/0.25 = 300 to 1 in power consumption is normal for CMOS circuits and can be even greater in some cases.

Now let V_{IN} be changed to something above +2.6 V. Transistor P1 turns off and N1 turns on. With N1 on, the gates of P2 and N2 are discharged and V2 goes high. With V2 high, an output current flows through P2 and R_L. Because of the direction of the arrow for I_O, the output current with V2 high is shown on CMOS data sheets as $-I_O$.

The amount of output current that flows depends on the magnitude of R_L and the turned-on resistance of P2. The output capability of typical 2-input NAND and NOR gates is shown in Fig. 2-8. With the output high and R_L connected to ground (Fig. 2-11), current flow will be low if R_L is large. As a result, the output voltage will be high and close to V_{DD}. As R_L is made smaller, the output current increases (from 0 toward −3 mA) along the lower

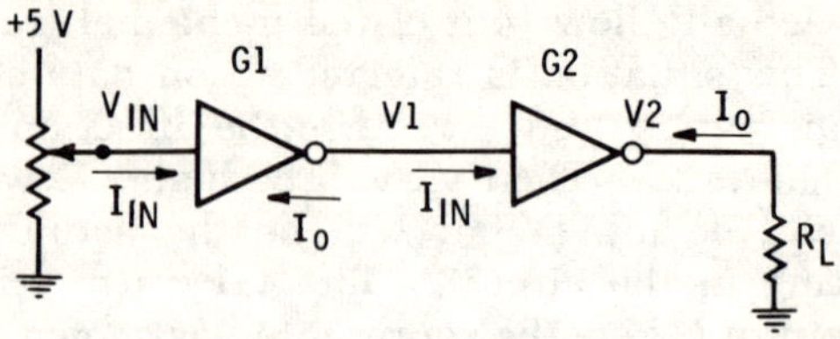

(A) Two-inverter logic.

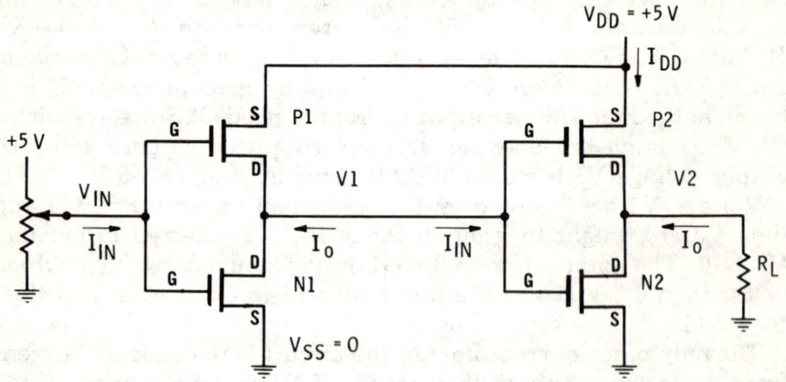

(B) Circuit.

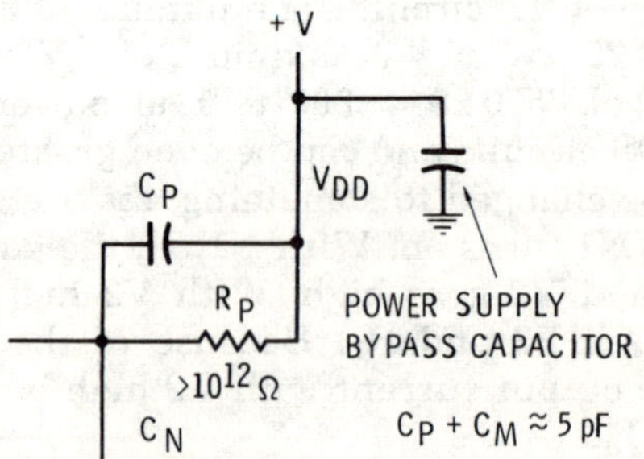

(C) Equivalent input circuit.

Fig. 2-10. Logic chains.

curve as shown at (B) for $V_{DD} = +5$ V. When $R_L = 0$, output voltage is 0 and output current is maximum. Even with $R_L = 0$, the output current in a typical CMOS gate is still less than 4 mA (for $V_{DD} = +5$ V). For proper operation of logic systems, the output voltage would not normally be allowed to fall much below 3.6 V. This is achieved when the circuit is confined to operating over the heavy part of the curve. For 3.6 V out, R_L should be 1.8K or larger, which represents an output current of about 2 mA. When a CMOS gate is driving another CMOS gate, the effective R_L becomes 10^{12} ohms or so, and thus maximum output voltage is obtained.

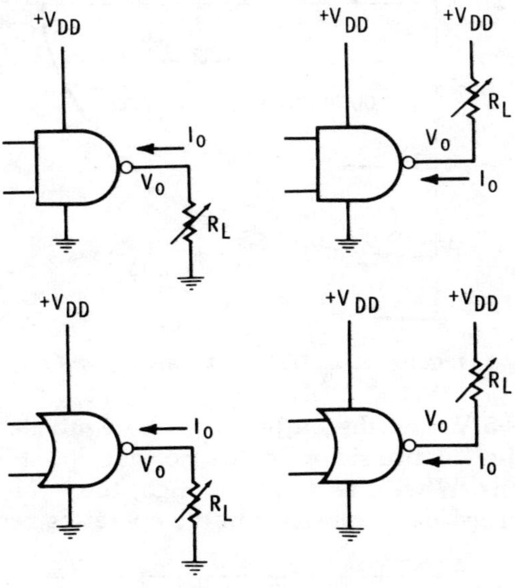

Fig. 2-11. Two-input NAND and NOR gate loads.

When the output is low, and R_L is connected to ground, no load current flows, and V_O is 0. But if R_L is connected to $+V_{DD}$, a low output allows current flow from V_{DD} through R_L, through one or more n-channel transistors to ground. The upper curve in Fig. 2-8B shows how output voltage varies with load current. The output voltage stays below 1.5 V if the current drawn is not more than about 3 mA. This is obtained as long as R_L is 500 ohms or larger.

The output characteristic varies as a function of V_{DD}. Operation at $V_{DD} = +10$ V and at +15 V is shown by the curves of Figs. 2-12 and 2-13. The output voltage-current curves represent output impedance, and the slope of the curve gives the resistance. Thus,

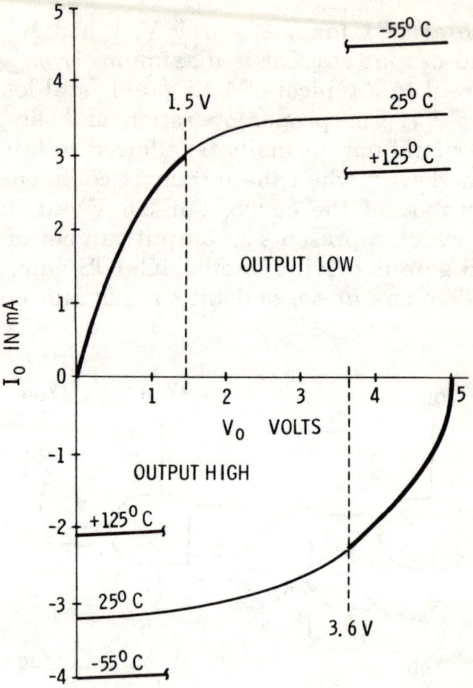

Fig. 2-12. Output curves for Fig. 2-11 with V_{DD} = +5 volts.

with V_{DD} = +5 V, and the output low, the equivalent resistance of the turned-on n transistor in the normal operating region is about 400 ohms. When the output is high, the equivalent resistance of the turned-on p transistor in the operating region is about

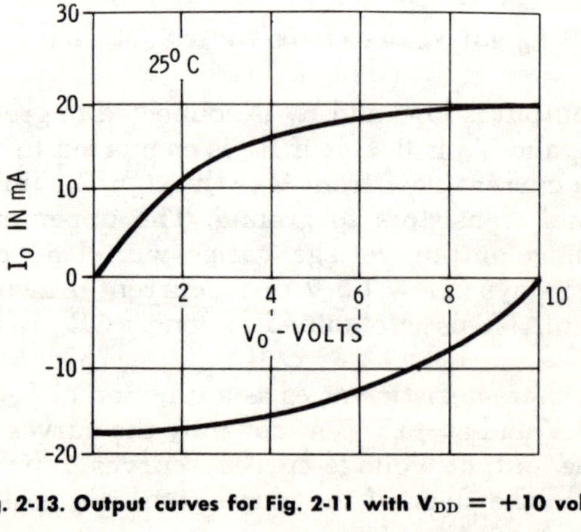

Fig. 2-13. Output curves for Fig. 2-11 with V_{DD} = +10 volts.

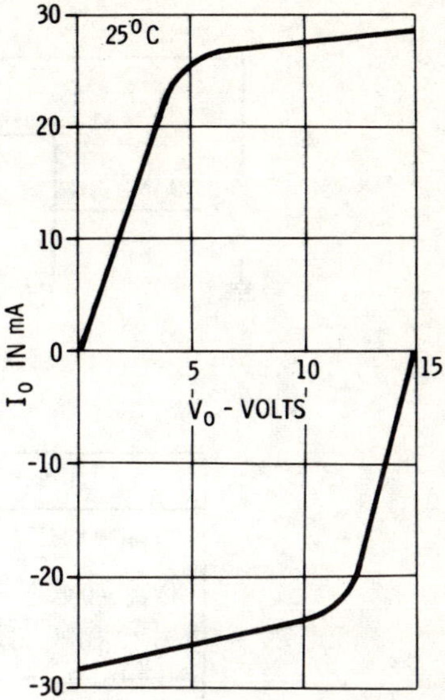

Fig. 2-14. Output curves for Fig. 2-11 with V_{DD} = +15 volts.

600 ohms. When V_{DD} is set at +10 V and +15 V, as in Figs. 2-13 and 2-14, the equivalent resistances change as indicated in Fig. 2-15. Note that the curves shown are for 2-input NAND and NOR gates. The information is usually presented in tabular form on data sheets instead of being presented as a curve, and in terms of voltage and current instead of resistance.

V_{DD}	OUTPUT Z WITH OUTPUT:	
	LOW	HIGH
+5 V	400 Ω	600 Ω
+10 V	166 Ω	250 Ω
+15 V	200 Ω	350 Ω

Fig. 2-15. Change of output impedance with change in V_{DD}.

For the 54C/74C family, the current sourcing and sinking capability for some gates is listed in terms of short-circuit current. Thus, for 54C00/74C00, the output short-circuit current range is shown in Fig. 2-16. A 5 to 1 range of current for individual devices is within acceptable limits.

The output resistance of CMOS is seen to be relatively high compared with that of bipolar circuits and to have relatively large variation from device to device. The practical effect of this is that CMOS gates should work into high-impedance loads.

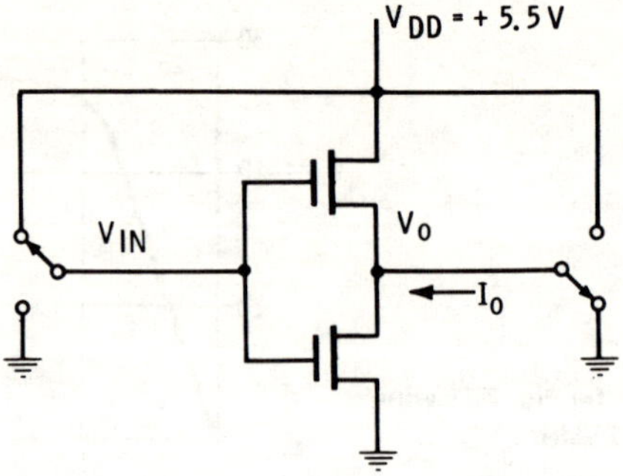

CONDITIONS	I_o (mA)
V_{DD} = + 5.5 V V_{IN} = 0 V V_o = 0 V	1 mA MIN 6 mA MAX
V_{DD} = 5.5 V V_{IN} = 5.5 V V_o = 5.5 V	1.5 mA MIN 10 mA MAX

Fig. 2-16. Short-circuit current for 54C/74C gates.

Since the input impedance of CMOS gates is extremely high, a CMOS input makes an ideal load for a CMOS output.

Fig. 2-17 shows two CMOS gates feeding other CMOS gates. Gates G1 and G3 are high, and G2 and G4 are low. What are the actual input and output currents and voltages of these gates and how many other inputs can G1 and G2 feed?

The p transistor of G1 is fully on; thus V1 is at its maximum. The n transistor of G2 is fully on; thus V2 is at its minimum. The final value of V1 and V2, then, depends on how much current the gates source or sink.

Consider G1 first. It is supplying inputs of G3, G4, G5, and X others. Each input represents a 5-pF capacitor in parallel with a 10^{12}-ohm resistor to ground. Ignore the capacitance for the moment and consider just steady-state dc conditions. The data sheet for a 2-input NAND circuit such as G1 (4000 series) states that if the current out of G1 is 0, V1 will be 4.99 V ($V_{DD} = +5$ V) for most devices at 25° C, and that at the worst temperature extreme, +125° C, it will be at least 4.95 V. With V1 = 4.95, the current into input B of G3 will be $4.95/10^{12} = 4.95$ pA. This much load current will not cause much of a drop across the P1 transistor, and V1 will remain close to +4.95 V. The situation for G4 is the same. Then how many gates can G1 drive before V1 drops to an unacceptable level?

To state the problem a little differently, how low can V1 fall before it is no longer a satisfactory high? Data sheets state that the minimum allowable value for a high input, under the worst temperature conditions and for $V_{DD} = +5$ V, is +3.6 V. Therefore, we can keep adding CMOS inputs until V1 sags to 3.6 V. Referring to Fig. 2-12, we see that the output voltage of a gate

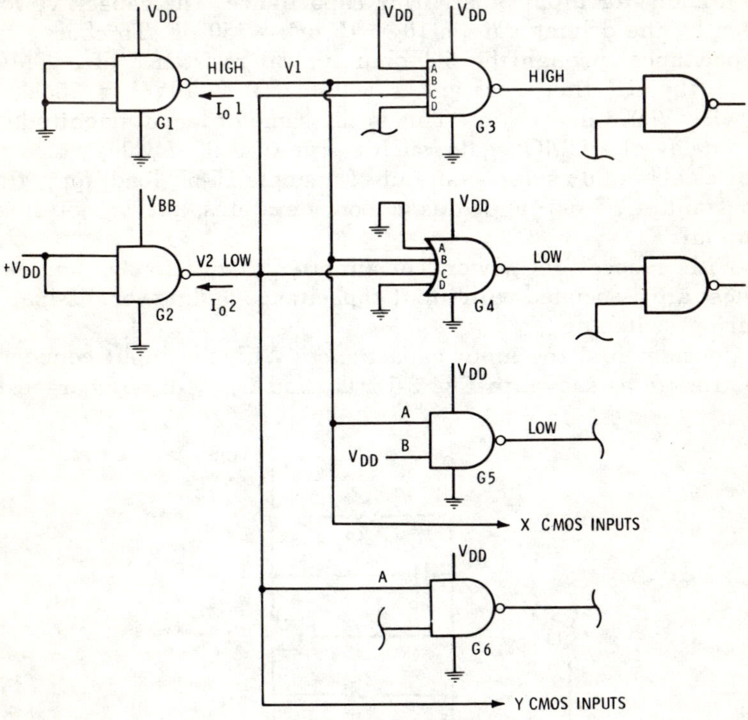

Fig. 2-17. CMOS fanout.

falls to 3.6 V when the output current is about 2 mA. With V1 = 3.6 V, the current drawn by any one input will be $3.6/10^{12} = 3.6$ pA. Thus, the number of inputs G1 can supply before V1 drops below 3.6 V is $2 \times 10^{-3} = N \times 3.6 \times 10^{-12}$, and $N = 555 \times 10^{6}$.

A similar situation exists for G2, where V2 is low. In this case we can assume that current flows from V_{DD} through the input structure of G3 and into the turned-on n transistor of G2 and to ground. We need to know the maximum voltage G3 requires at an input to accept it as a true low. From the data sheet, we find that V2 can rise to +1.4 V (worst-case temperature) and G3 will still respond properly.

From Fig. 2-12, we find that with V2 = 1.4 V, a current of about 3 mA can flow. Thus, if V2 = 1.4 and V_{DD} = 5 V, the voltage drop from V_{DD} through the gate structure of G3 to V2 is 5 − 1.4 = 3.6 V. How much current flowing through 10^{12} ohms gives 3.6 V? Fan-out = 833×10^6.

Based on purely dc considerations, fan-out is essentially infinite. Consider, instead, a fan-out of 10 and allow 15 pF for each input, to account for input plus wiring capacitance. The capacitive load seen by the driving gate is 10 × 15 pF = 150 pF. To charge this capacitance through the 500-ohm typical resistance of a CMOS gate, the RC time constant is $500 \times 150 \times 10^{-12} = 75{,}000 \times 10^{-12} = 0.075\ \mu s$ = 75 ns. This is the same order of magnitude as the delay of a CMOS gate, such as type 54C00/74C00, which requires about 20 ns (at a fan-out of a single 15-pF load, for a time constant of 7.5 ns), and thus ac considerations, not dc, determine fan-out.

With respect to power consumption, two effects dominate. These are associated with input capacitance and power dissipation during switching.

Consider first the input capacitance. With the input connected to ground, as shown in Fig. 2-18, C_P and C_{IO} will be charged up

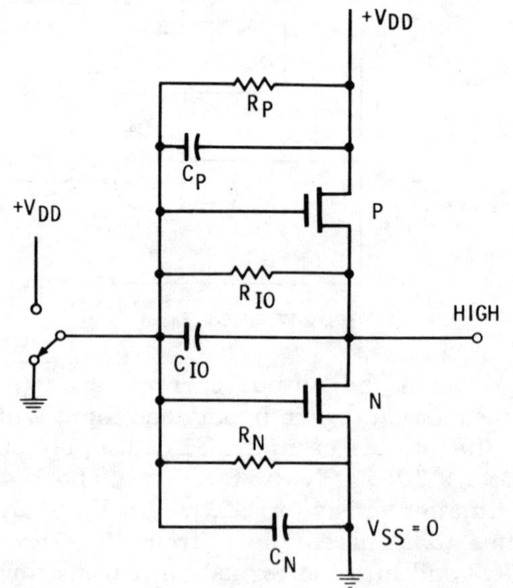

Fig. 2-18. CMOS inverter capacitance and leakage resistance.

to V_{DD}, and C_N will have both sides connected to ground and will have no charge. When the input is switched from 0 V to $+V_{DD}$, C_N has to be charged, and C_P becomes discharged. At the same time, C_{IO} becomes charged in the opposite direction. (In addition, each capacitance of the input protective diodes contributes to the total capacitance of C_P and C_N.) The net result of all these capacitances acting together is lumped into an equivalent input capacitance of about 5 pF. Every time the input is switched to $+V_{DD}$, this input capacitance draws an exponential charging current until it is fully charged. When the input is switched to ground, the input capacitance is discharged. Thus, a spike of current is drawn from the power supply every time the circuit changes state. The more often the input changes, the more the average power drawn by the circuit increases.

Power consumption, then, consists of dc power plus dynamic power and is given by:

$$P = P_q + CfV^2$$

where,

P_q is the quiescent dissipation and is very small,
C is the load capacitance that must be charged,
f is the frequency of operation in pulses or clocks per second,
V is the supply voltage.

Fig. 2-19 shows how power consumption varies as a function of both operating frequency and power-supply voltage.

A second type of power consumption is caused by the rise and fall times of the input signals. In Fig. 2-20A, when V_{IN} is near 0,

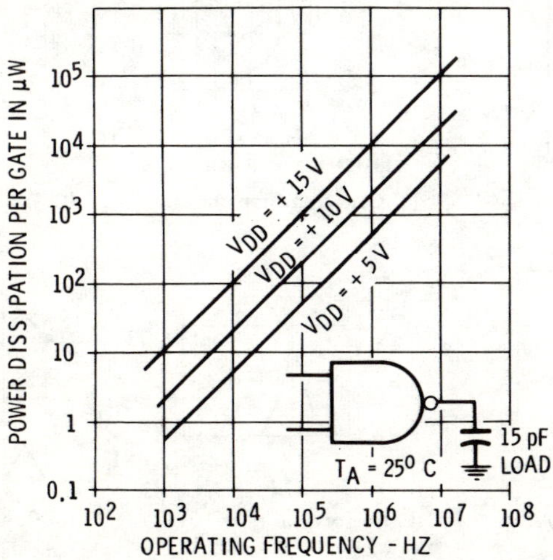

Fig. 2-19. Power consumption of CMOS gates.

the p transistor is fully on and the n transistor is fully off. As V_{IN} is slowly increased, the threshold of the n transistor is passed (see Fig. 2-20B) and the n transistor begins to turn on. Since the p transistor is also on at this time, current can flow from V_{DD} through the transistors to ground. This current takes the form shown in Fig. 2-20C and reaches a peak of about 9 mA under worst conditions with $V_{DD} = +15$ V.

When the peak current is flowing, the circuit is using (and wasting) about $9 \times 10^{-3} \times 15 = 135$ mW. Type 54C04/74C04 is a hex inverter with a maximum power rating of 500 mW for the package. If all 6 gates in such a package were operating at the 9-mA point, the package dissipation would be $135 \times 6 = 810$ mW. If the condition lasted long enough, the circuit would overheat and be destroyed. If only 2 or 3 gates in the same package were operating at this point, the package might not overheat and the circuit might continue to function.

When V_{DD} is +10 V, the effect (shown in Figs. 2-20D and E) is not as severe as at +15 V, and it becomes even less of a problem at $V_{DD} = +5$ V, where the peak current is very small.

A current spike is drawn from the supply whenever V_{IN} changes from low to high or high to low (Fig. 2-20F). The width of the current spike is a function of how fast V_{IN} changes. The

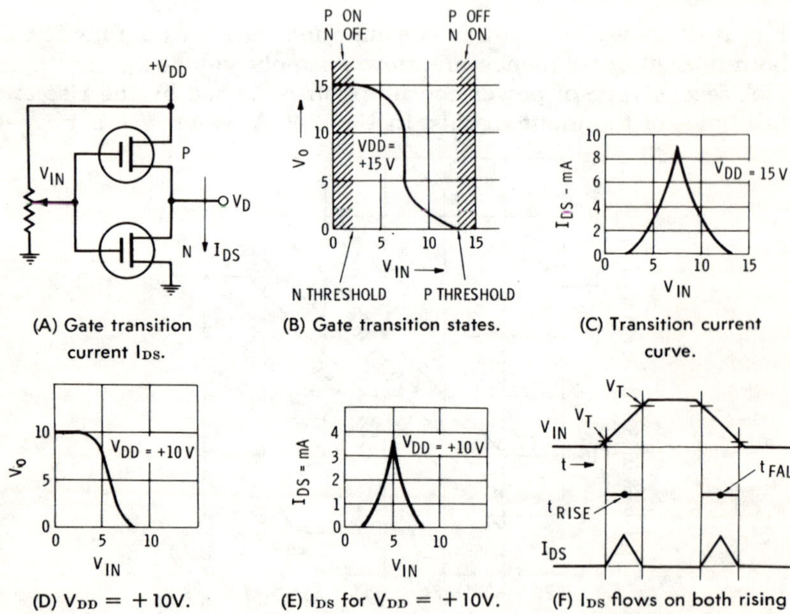

(A) Gate transition current I_{DS}.

(B) Gate transition states.

(C) Transition current curve.

(D) $V_{DD} = +10$V.

(E) I_{DS} for $V_{DD} = +10$V.

(F) I_{DS} flows on both rising and falling inputs.

Fig. 2-20. Transition current in CMOS gate.

longer I_{DS} remains above 0, the more power the circuit dissipates. The power dissipation during switching, P_{VI}, can be calculated from $P_{VI} = [(V_{DD} - 2\ V_T)/2] \times I_{DS}$ max (t-rise + t-fall) $\times$ f.

In actual practice, P_{VI} is measured by operating the circuit at no load and various frequencies for the common V_{DD} voltages. Most data sheets do not list a value for P_{VI}. When circuits are of the type that cannot exceed their dissipation limits for any combination of inputs and V_{DD}, a value for P_{VI} is not really necessary. In other cases, manufacturers will list minimum rise and fall times for inputs. These limits may be required for either proper circuit operation, power dissipation, or both.

In a CMOS family recently developed by Fairchild Semiconductor Corp., the transfer characteristic has been squared up as indicated in Fig. 2-21A. This has been accomplished by using small-geometry input transistors and large-geometry output transistors for low output impedance. A 2-input NAND gate in this family (34000) is shown in Fig. 2-21B. Although the gate uses eight transistors instead of four, it is faster than regular 4000 and 4500 CMOS and uses the same pinouts and part numbers as 4000 ICs. (Some devices in this family, those with part numbers 340XXX, duplicate devices in the 54C/74C family.)

DEVICE PROTECTION

MOS transistors in ICs can easily be destroyed by voltage transients. If the device is n channel and is to be operated with +15 V on the drain and 0 V on the source, then the gate will normally vary between 0 V and +15 V. When the gate is at 0 V, the gate and source are at the same potential and the electrical field applied to the insulation between them is 0. In the region between drain and gate, however, a potential difference of 15 V exists across the insulation.

The thickness of insulation between gate and substrate ranges from about 1000 to 1500 angstroms. Since one angstrom is 10^{-8} cm, the electric field strength in a 1200-angstrom-thick insulation between gate and drain is $15/(1200 \times 10^{-8}) = 1.25 \times 10^6$ volts/cm. For comparison, air at atmospheric pressure breaks down (sparks over) at about 31×10^3 V/cm (for a standard 1-cm gap).

When a person walks across a waxed floor, he can generate a static charge of up to 15 kV on his body. What happens when a person charged to 15 kV touches the gate of a MOS transistor? Instantaneously, the charge is shared between the person and the gate structure, and the sharing is in proportion to their respective capacitances. Since a person has a capacitance of about 300 pF and the gate capacitance is about 5 pF, the gate will not be able

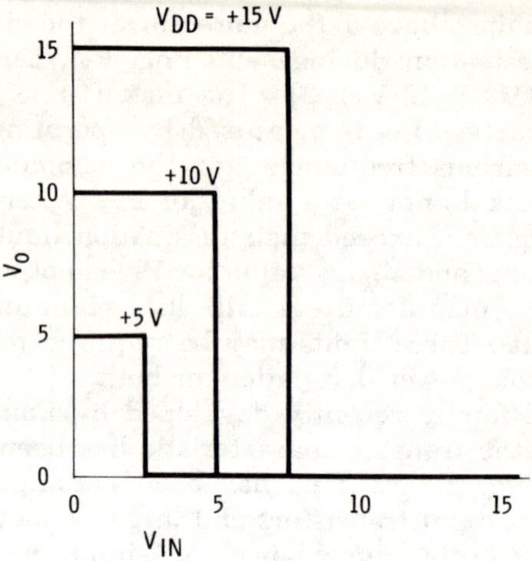

(A) Transfer curves.

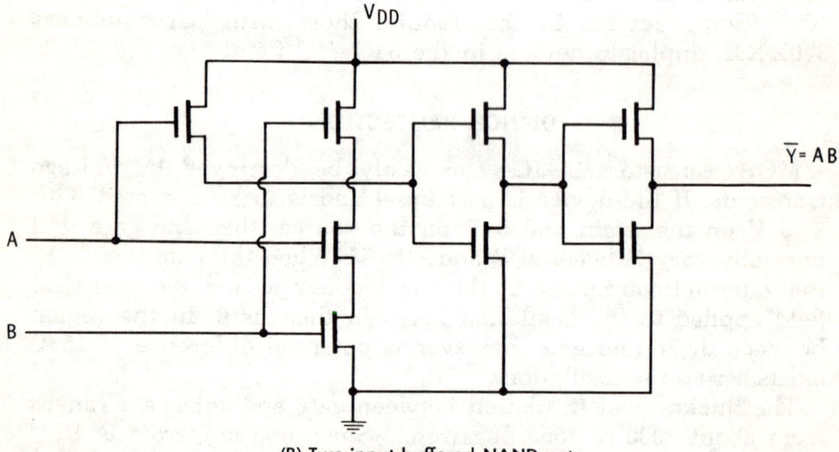

(B) Two-input buffered NAND gate.

Fig. 2-21. Buffered CMOS family.

to accept much charge but instead will rise nearly to the same voltage as the person. When this happens, the electric field from gate to source, drain, and substrate rises 1000 times higher than the normal 1.25×10^6 V/cm. Not even the gate insulation can withstand this, and the device is destroyed in the first few nanoseconds after contact occurs.

To protect MOS devices against stray fields of the type indicated above, manufacturers add protective networks at all MOS inputs. It is at the gate input that destructive transients can enter, usually not at the source or drain.

Input protection networks take the forms shown in Fig. 2-22. Consider input A of the 54C/74C device in Fig. 2-22A. If A attempts to go more positive than V_{DD}, diode D1 conducts and clamps the input signal on A to $+V_{DD}$. If A attempts to go more negative than V_{SS}, diode D2 conducts and A is clamped to V_{SS}. Actually, since the diode must be forward-biased to conduct, A can go about 0.5 V more positive than V_{DD} and about 0.5 V more negative than V_{SS}. This small excursion can easily be tolerated.

Resistance R1, as shown in Fig. 2-22B (another input circuit used on some 54C/74C ICs), in series with the gate prevents fast-rising signals from charging the gate before the diodes have a chance to conduct. In combination with the stray capacitance of D1 and D2, R1 acts to slow the application of fast-rising or fast-falling inputs to the gate. The time constant of the protection circuit is much less than the time constant of the gate structure itself, and thus the overall speed of the logic element is not affected.

Input protection for the 4000 series of devices is shown in Fig. 2-22C. The extra diode provides another level of protection and serves as a backup in case the first diode should be destroyed. Similar protection circuits are used in all CMOS and most MOS families. A few MOS ICs do not have protection networks on inputs and these devices require special handling.

Fig. 2-23A shows the cross section of a CMOS circuit. The substrate is shown as n material with a p-channel transistor and a p island with an n-channel transistor. To make connections to the gate, source, and drain, metal pads are laid down. The IC is seen to consist of a number of pn junctions. The polarity of the voltages applied to the circuit must at all times be such that no pn junction is ever forward-biased into conduction, except for the protective diodes, which are not shown. If forward bias into conduction does occur, transistor action stops and the device will usually be destroyed.

Three groups of pn junctions exist in the circuit: p well to n transistor, p well to n substrate, and p transistor to n substrate. The integrity of the pn junctions can be maintained, provided the following relationships are held at all times:

1. The n substrate must always be more positive than the p well and the p-channel transistor.
2. The voltage applied to the n-channel transistor must always be more positive than the voltage applied to the p well (or p substrate).

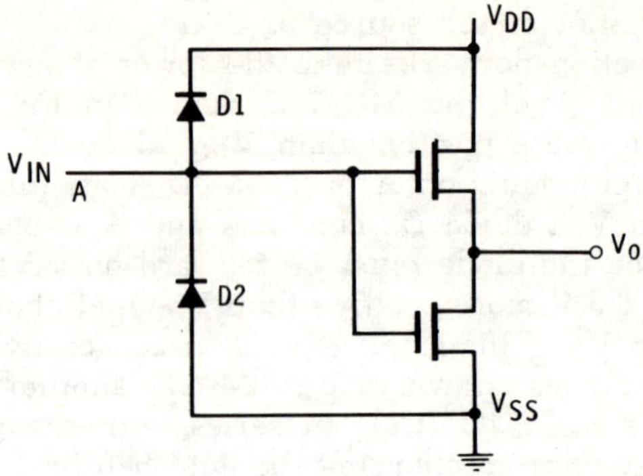

(A) Diode protection of 54C/74C input.

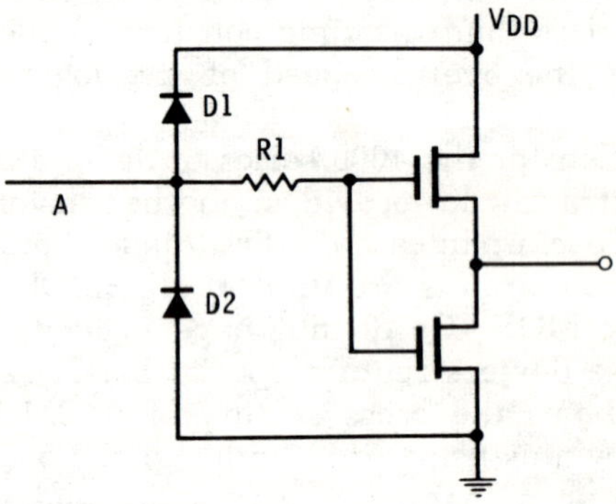

(B) Diode and resistor protection of 54C/74C input.

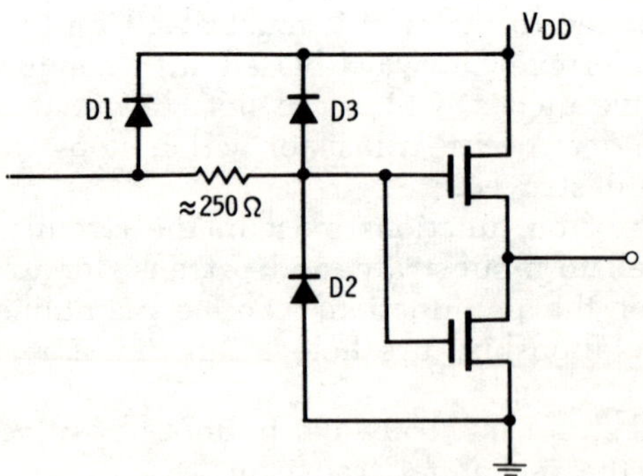

(C) Protection of a 4000-series input.

Fig. 2-22. CMOS input protection.

Suppose that instead of using a p well in an n substrate, an n well is used in a p substrate. Then the circuit will look as shown in Fig. 2-23B. The pn junctions occur in the same relationships as before.

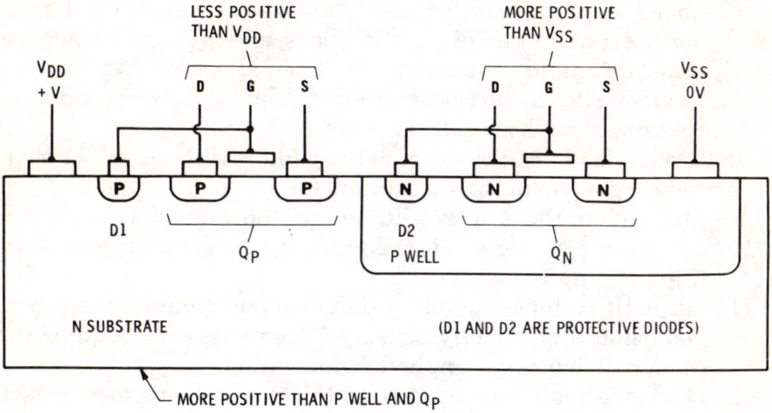

(A) P well in n substrate.

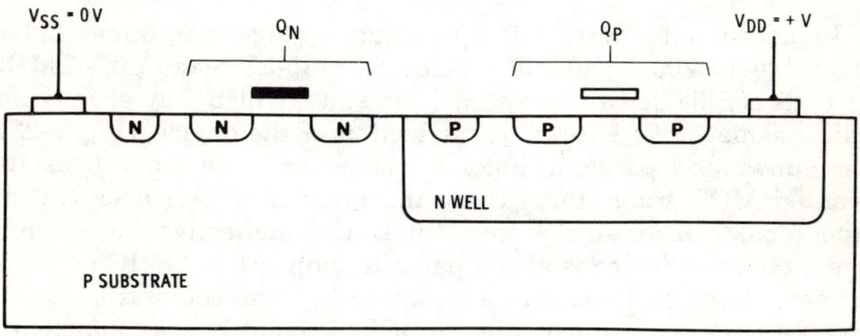

(B) N well in p substrate.

Fig. 2-23. Cross section of CMOS structure.

Even though CMOS and most MOS inputs have protective networks, the energy these circuits can absorb is limited. If too large a transient takes out the protective network, the next transient may take out the circuit. To prevent this from occurring, the following guidelines should be observed:

1. Soldering iron tips should be grounded. Alternatively, use IC sockets and make all connections while the device is not inserted.
2. Test equipment should be grounded.

3. Automated handling equipment should be conductive and grounded.
4. Metal or conductive plastic work surfaces should be used and should be connected to ground.
5. Persons handling the devices should be grounded.
6. Shoes, clothing, and smocks should be conductive if possible. Certain types of plastic shoes and nylon clothing can generate static electricity and should be avoided.
7. Devices should not be stored or carried in nonconductive containers such as plastic trays.
8. Device leads can be shorted together with metal carriers, conductive foam, or aluminum foil.
9. The higher the relative humidity, the better.
10. Ionized-air blowers can reduce and prevent charge buildup if grounding is not practical.
11. All limit ratings, including voltages and timing when specified, should be strictly observed. Some devices require that certain voltages be applied before others.
12. Devices should be inserted and removed from sockets with all voltages off. All unused inputs should be connected to V_{DD}, or V_{SS}, or other inputs.

In addition to destruction by high voltage—anything higher than 100 to 200 V, and far below the 15-kV example—CMOS circuits can be driven into a latch-up state, which may or may not cause damage. Note in the cross section of the device (Fig. 2-23), that unwanted parasitic bipolar transistors exist as well as the wanted MOS transistors. Thus, the n-channel device is also an npn bipolar transistor with a thick and ineffective base, while the p-channel device is also a parasitic pnp. Also, the IC contains a vertical npn and a lateral pnp, which are connected as shown in Fig. 2-24. The circuit acts like an SCR. If point X goes sufficiently positive with respect to V_{DD}, Q1 will turn on, which will turn on Q2, connecting V_{DD} to V_{SS} through the p-substrate resistance. The circuit remains latched until V_{DD} is reduced below the holding current of Q2, or until the IC burns up.

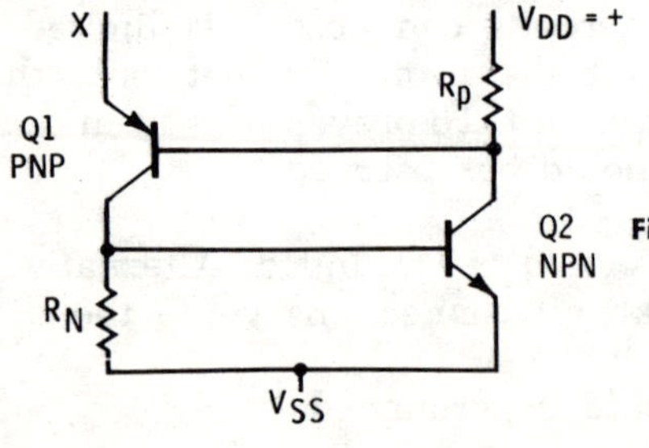

Fig. 2-24. Parasitic CMOS latch.

3

CMOS Circuit Applications

Current flow in an n-channel MOS transistor is through n-type material only and not through any pn junction. The circuit is purely resistive, or ohmic, and follows Ohm's law as long as the gate-to-source voltage is high enough to turn the device on. The same considerations hold for p-channel MOS transistors.

CMOS TRANSMISSION GATES

A transmission gate uses the resistance-only characteristic of MOS transistors to form a bilateral switch. The switch conducts in both directions and is formed by connecting an n- and a p-channel transistor as shown in Fig. 3-1A. With the gate of Q_N at $-V_{SS}$, Q_N can never turn on, because its source will always be more positive than the gate, no matter whether V_{IN} is at $+V_{DD}$, $-V_{SS}$, or somewhere in between. At the same time, the gate of Q_P is positive and can never go more negative than its source. Thus, Q_N and Q_P are both in the high-impedance state and output voltage V_O remains close to 0.

Now change the gate bias to that shown at Fig. 3-1B. As long as V_{IN} is less than V_{DD}, Q_N will be on. As V_{IN} approaches $+V_{DD}$, Q_N begins to turn off and its impedance rises. At the same time, however, Q_P is fully on and is a low impedance; thus, V_O rises toward V_{IN}. As V_{IN} approaches $-V_{SS}$ (Fig. 3-1C), Q_P begins to turn off while Q_N is fully on and V_O approaches V_{IN} or $-V_{SS}$. When V_{IN} is between $+V_{DD}$ and $-V_{SS}$, and not close to either, both Q_N and Q_P are on. These three states of circuit operation are indicated in Fig. 3-1D, where region I refers to operation with V_{IN} near $+V_{DD}$, region II refers to operation with V_{IN} near $-V_{SS}$, and region III refers to operation in between. The impedance of the transmission gate is the parallel resistance of the two transistors as V_{IN} varies between $-V_{SS}$ and $+V_{DD}$.

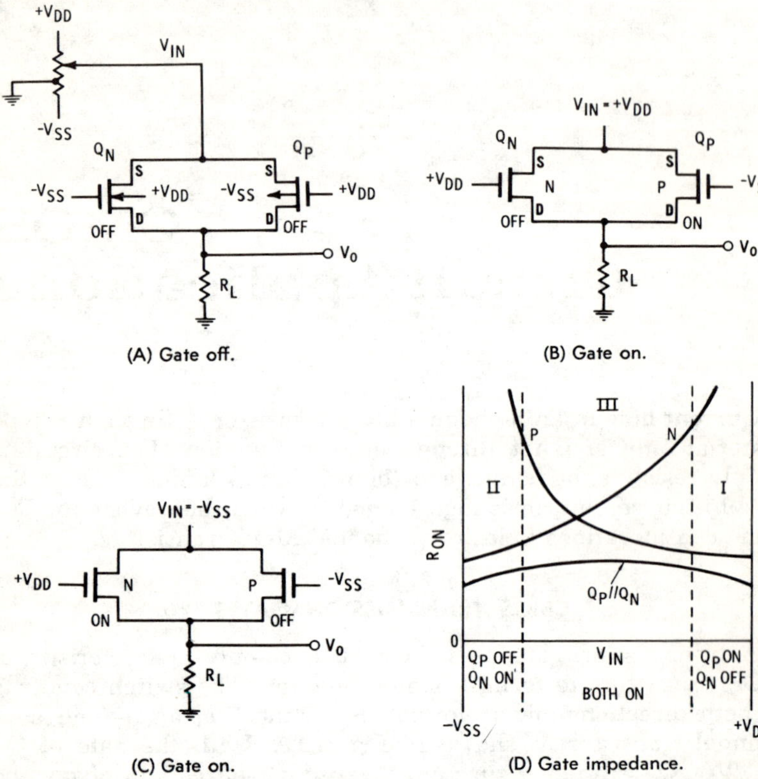

Fig. 3-1. Operation of CMOS transmission gate.

Because both a positive and a negative signal are required for proper operation of the transmission gate, the control input is applied to an inverter, as in Fig. 3-2A, which gives both A and $\overline{A}$, where $A = +V_{DD}$ and $\overline{A} = -V_{SS}$. The signal the transmission gate can handle is limited to values between $-V_{SS}$ and $+V_{DD}$, since signals of greater magnitude would cause one or more pn junctions to become forward-biased, as indicated in Fig. 3-2B. For an ac signal with peak amplitudes of +5 and −5 V, V_{DD} can be set at +5 V or more, and V_{SS} at −5 V or lower.

Transmission gates are used in CMOS logic circuits as switches, where they are usually shown in one of the ways indi-

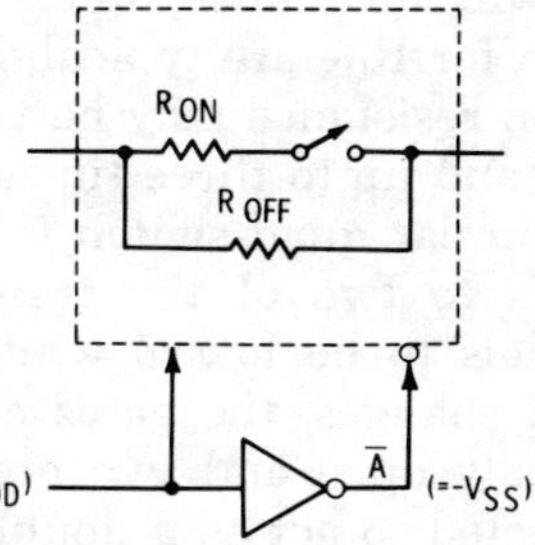

(A) Equivalent circuit of transmission gate.

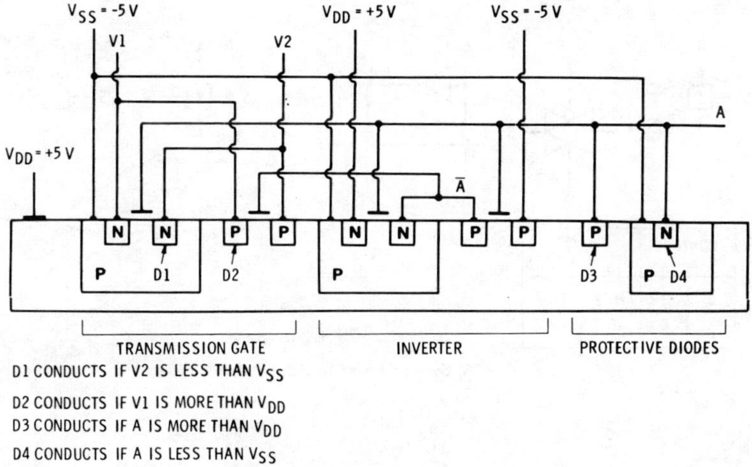

(B) Cross section of transmission gate.

Fig. 3-2. Transmission gate circuits.

cated in Fig. 3-3A, with S the signal and A and $\overline{A}$ the control. The transmission gate is designed to operate as a break-before-make circuit. This allows two or more signals to feed the same load, as in Fig. 3-3B, without being connected together by two gates being on simultaneously from the same control signal. Break-before-make is obtained by designing the transistors and associated capacitances so that turn-off occurs faster than turn-on.

Transmission gates are also useful in analog circuits, where they are used to switch analog signals such as audio channels electronically. Type MM4616A/5616A and type 4016 are quad, bilateral switches that can be used with analog signals having a total excursion of less than 15 volts. Fig. 3-4 shows the circuit of the IC, and Fig. 3-5 shows the variation of on resistance as a function of input signal, load resistance, V_{DD}, and V_{SS}. As long as R_L (Fig. 3-4B) is sufficiently larger than R_{ON}, the variation in

R_{ON} will not cause much distortion in the signal. The values shown for R_{ON} are typical, however, and are applicable for 25°C. The on resistance may be twice as much for individual devices at 25°C, and up to three times as much at 125°C.

A similar quad switch is type 4416, which reverses the control signals to two of the transmission gates and thus causes two channels to be low Z while two are high Z. When the control signal changes, the gates change, giving the same effect as two normally open and two normally closed switches. These can be connected to act as a double-pole, double-throw switch.

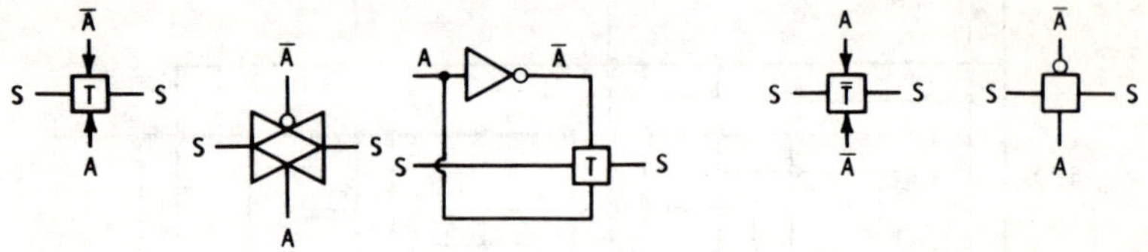

CONTROL		GATE	
A	$\bar{A}$	T	$\bar{T}$
0	1	HI Z	LO Z
1	0	LO Z	HI Z

(A) Symbols and logic.

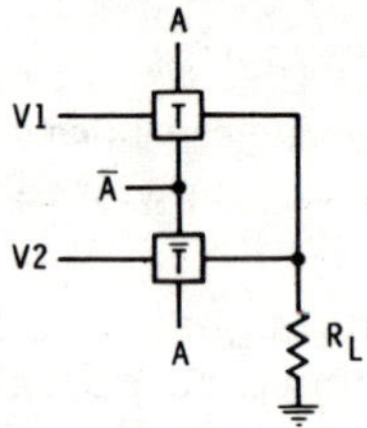

(B) Break-before-make switch.

Fig. 3-3. Transmission gate symbols and use.

THREE-STATE CMOS OUTPUTS

In many data transfer applications it is desirable to be able to feed one common set of wires from a number of different points, as in Fig. 3-6A. In such a system, one terminal can talk while the others listen. The active push-pull circuit of the standard CMOS device does not allow such operation.

As shown in Fig. 3-6B, if Y1 is high and Y2 is low, a direct path exists from V_{CC} (same as V_{DD}) through P1 and N2 to ground. Since the impedances of the transistors are relatively high and nearly equal, the circuit may not be destroyed, but V_O

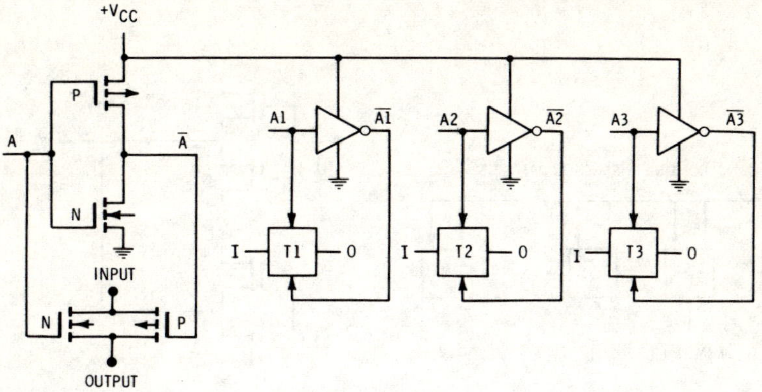

(A) Circuit.

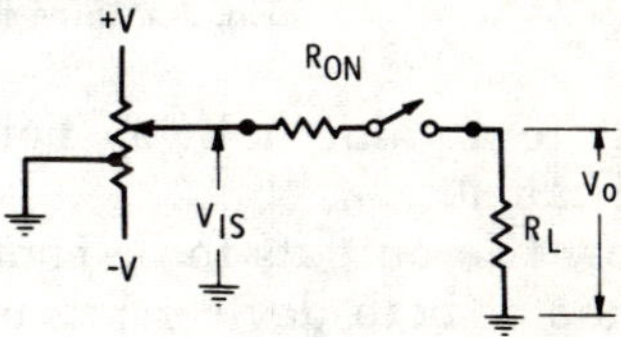

(B) Equivalent circuit, each switch.

$$V_O = V_{IS} \times \frac{R_L}{R_L + R_{ON}}$$

Fig. 3-4. Quad bilateral switch combination.

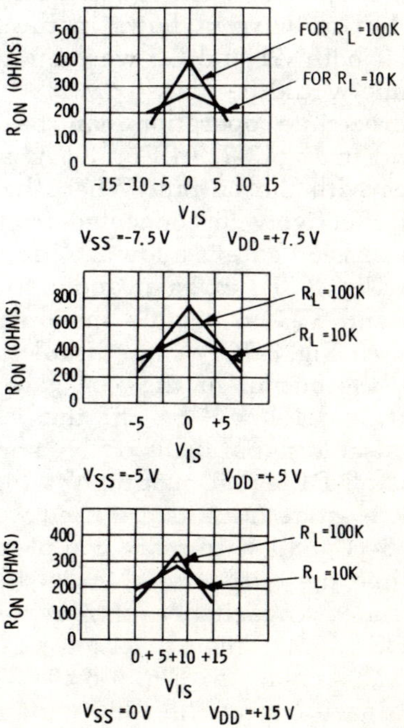

Fig. 3-5. Curves showing variation of on resistance for Fig. 3-4.

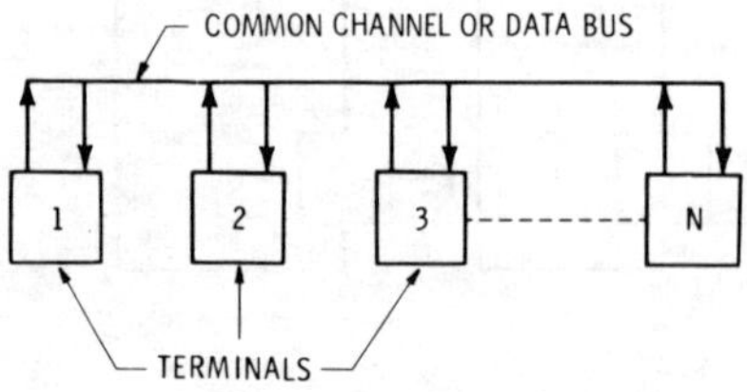

(A) Common-data-bus arrangement.

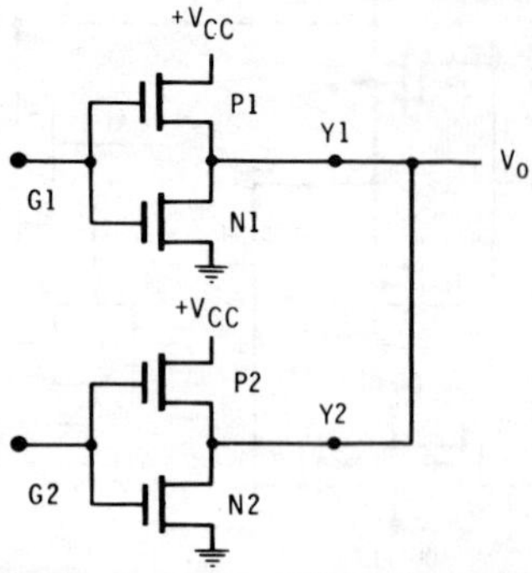

(B) CMOS outputs cannot be tied together.

Fig. 3-6. Three-state outputs.

will settle to a value halfway between V_{CC} and V_{SS} and will be neither 1 nor 0.

To allow two outputs to be connected together as in Fig. 3-6B, the outputs of both devices are modified so that they are either high, or low, or off, which gives three states for the circuit. When G1 is off, both P1 and N1 will be off and output Y1 will be a high impedance both to ground and to V_{CC}—it will be floating. If G2 is on when G1 is off, V_O will follow Y2. Output Y1 also follows Y2, but no current flows in the output transistors of G1. Note that if both G1 and G2 were on at the same time, the three-state scheme would be defeated.

Three-state operation can be achieved in several ways, as shown in Fig. 3-7. In Fig. 3-7A, a transmission gate is placed in series with the output. When the transmission gate is a high Z, Y is effectively disconnected from the output circuit. When the transmission gate is a low Z, Y follows the inverter.

In Fig. 3-7B, extra p and n transistors effectively disconnect V_{DD} and V_{CC} to disable the output when the disable signal is high. At Fig. 3-7C, a high disable signal turns off P_X to disconnect V_{DD}; the output of the NOR gate will be low, regardless of the input, and N1 will be off; thus, Y is a high Z. At Fig. 3-7D, a low disable signal turns off N_X and, acting through the NAND gate, turns off P1, again making Y a high Z.

Three-state outputs are used with simple gates, as indicated in Fig. 3-7E, and with more complex circuits.

When the output of a three-state device is disabled, it has an associated capacitance similar to the capacitance of a typical CMOS input. This capacitance ranges from 10 to 15 pF. Therefore, any device in Fig. 3-6 that is driving the bus must charge the capacitances of all the inputs and outputs that are connected, as well as all the interface wiring capacitance. This capacitance slows the speed of operation. In addition to the delay caused by the capacitance, allowance must be made for the time the disabling circuit requires to turn on and off.

The use of three-state outputs raises the possibility of all drivers being in the high-Z state at the same time. If this happens, the inputs of all circuits connected to the bus will be floating. The inputs would then follow any stray field, and this could result in the ICs being destroyed. To prevent this from occurring, pull-up and pull-down resistors can be connected to the bus. Alternatively, logic circuits can be used to detect the condition of all drivers off, and when this occurs, transmission gates can be used to connect pull-up and pull-down resistors.

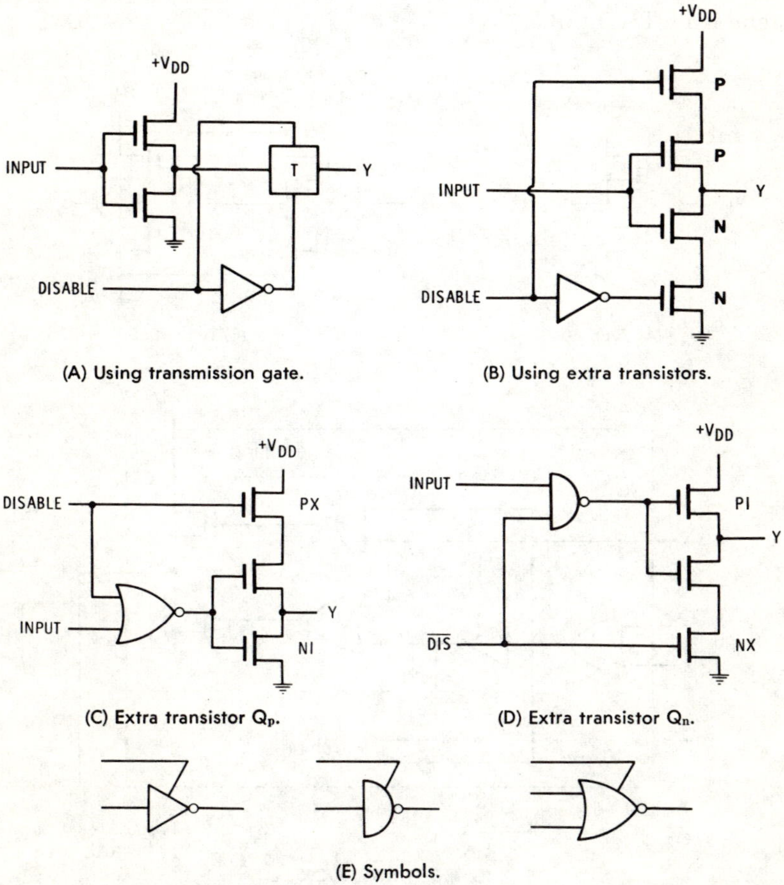

Fig. 3-7. Three-state control.

CMOS BUFFER OUTPUTS

Since most CMOS devices have a relatively high output impedance, and thus a limited ability to either source or sink current, a special output circuit having low impedance has been developed. Type 4049 (Fig. 3-8A) is a hex (six) inverter buffer, type 4050 (Fig. 3-8B) is a hex noninverting buffer but with an extra stage of delay, and type 70C95/80C95 (Fig. 3-8C) is a three-state noninverting hex buffer.

Circuit operation is in the manner of standard CMOS. To obtain a higher output current, the output transistors are made physically large compared with those in standard CMOS. As a result, each buffer can drive 11 or 13 (depending on the impedance rating of the specific device) low-power TTL inputs or one standard TTL input.

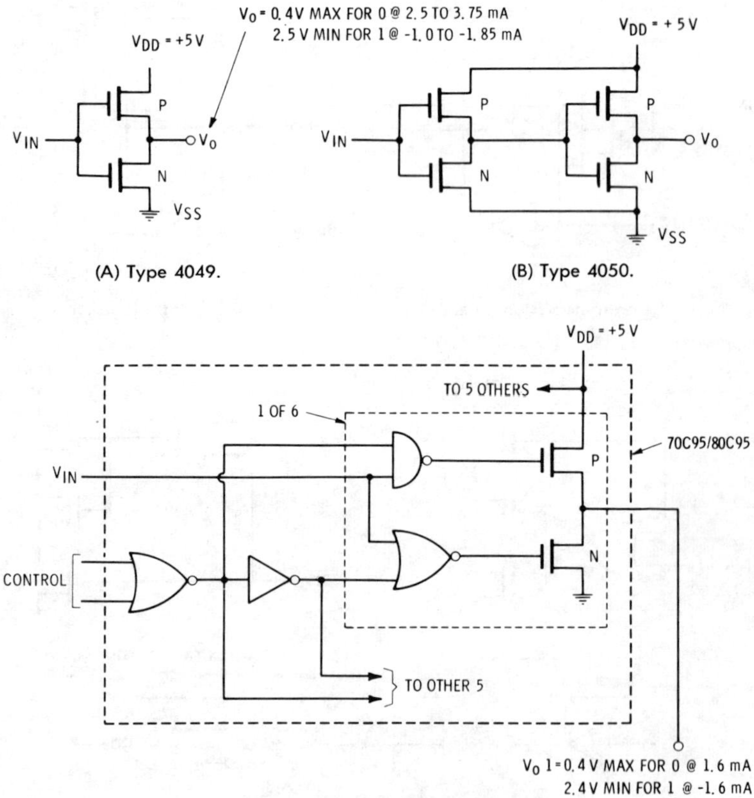

(A) Type 4049.

(B) Type 4050.

(C) Type 70C99/80C95.

Fig. 3-8. CMOS buffers.

When the ICs are operated with $V_{DD} = +10$ V, they can handle about four times as much current as at +5 V. A buffer output circuit may be added to more complex CMOS ICs whenever the extra drive capability is useful and the silicon real-estate situation allows it.

FLIP-FLOPS IN CMOS

A D-type CMOS flip-flop is shown in Fig. 3-9. The IC, type 54C/74C74, is a dual D-type flip-flop, and each circuit uses four NAND gates, six transmission gates, and four buffers. Consider first the operation of the transmission gates. When the clock signal is low, transmission gate T (Fig. 3-10) has the right control inputs for conduction and presents a low impedance to signal S in either direction, while the $\overline{T}$ gate has the wrong control inputs for conduction and blocks signal S in both directions.

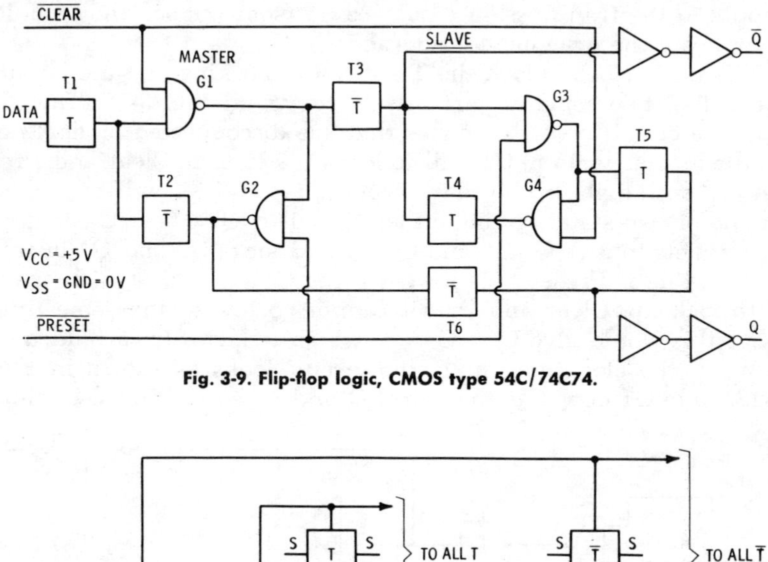

Fig. 3-9. Flip-flop logic, CMOS type 54C/74C74.

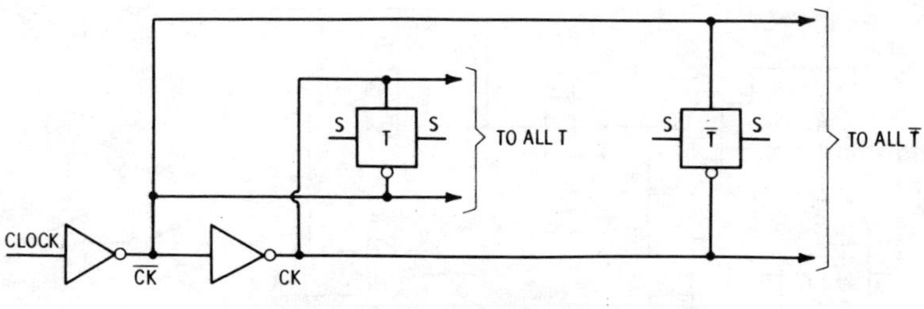

CLOCK	T (T1, T4, T5)	$\overline{T}$ (T2, T3, T6)
L	LOW Z	HIGH Z
H	HIGH Z	LOW Z

Fig. 3-10. Transmission gate operation of Fig. 3-9.

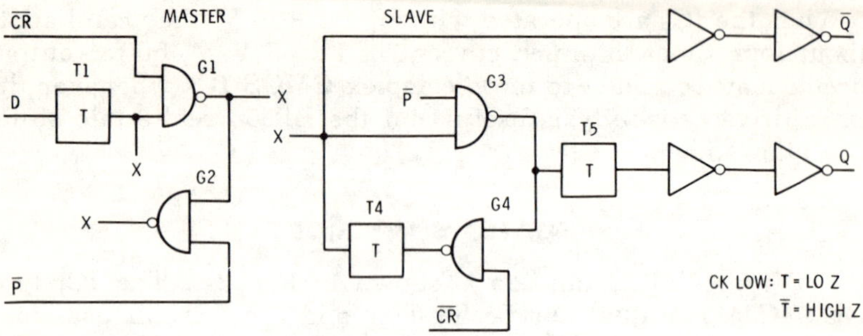

Fig. 3-11. Circuit of Fig. 3-9 with clock low.

When the clock is high, the control signals are reversed and now T blocks S while $\overline{T}$ conducts. Gates T and $\overline{T}$ thus act as switches, with T closed when the clock is low and $\overline{T}$ closed when the clock is high. In the logic diagram of the flip-flop, the control inputs to the transmission gates are not shown since they would complicate the drawing considerably.

When the clock is low, the $\overline{T}$ gates are blocking and the circuit looks like two separate circuits, as shown in Fig. 3-11. The Xs indicate circuit points or nodes that are disconnected or floating. In the Master section, G1 will follow D as long as Clear and Preset are both high. In the Slave section, G3 and G4 will be locked in one of two states, giving either $Q = 1$ or $Q = 0$. If Clear goes low, it acts to give $Q = 0$ and $Q = 1$ and locks G3 and G4 in this state, while if Preset goes low, it gives $Q = 1$ and $\overline{Q} = 0$, again with lock up. Clear and Preset cannot go low at the same time since this would give $Q = \overline{Q} = 1$, which is logically forbidden.

When the clock goes high, the circuit looks as shown in Fig. 3-12. Gate G1 now acts to control $\overline{Q}$ and to set G3 and G4, while

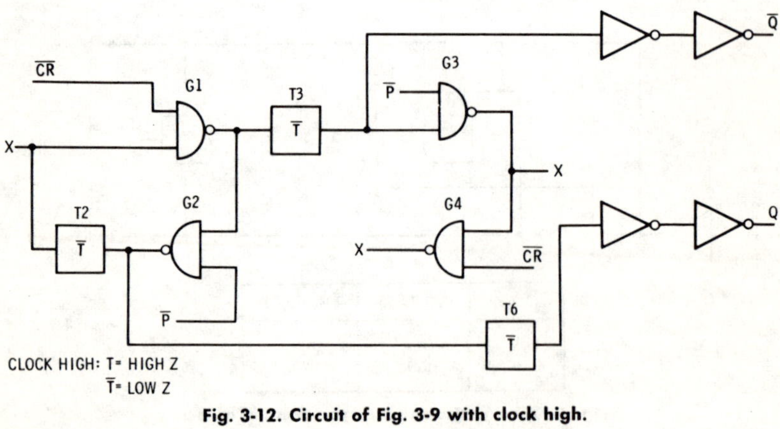

Fig. 3-12. Circuit of Fig. 3-9 with clock high.

CK	P	C	Q	$\overline{Q}$
X	0	0	1	1
X	0	1	1	0
X	1	0	0	1
(SEE B)	1	1	Q	$\overline{Q}$

(A) Clear and preset.

P	C	CK	D	Q
1	1	0	X	Q
1	1	1	X	Q
1	1	↑	D	D
1	1	↓	X	Q

(B) Clock control.

D	Qn + 1
0	0
1	1

Qn + 1 = PERIOD AFTER POSITIVE-GOING CLOCK TRANSITION.

(C) Flip-flop operation.

Fig. 3-13. Truth tables for Fig. 3-9.

G2 controls Q; input D has been locked out, but Clear and Preset will set the circuit if either goes low.

Circuit operation is summarized in the truth tables in Figs. 3-13A, B, and C.

The D-type flip-flop 4013 (Fig. 3-14) is similar in operation to the 54C74/74C74 but uses NOR gates instead of NAND gates and not as many of them. Instead of Preset and Clear inputs being high during normal operation, the 4013 requires that these inputs be low.

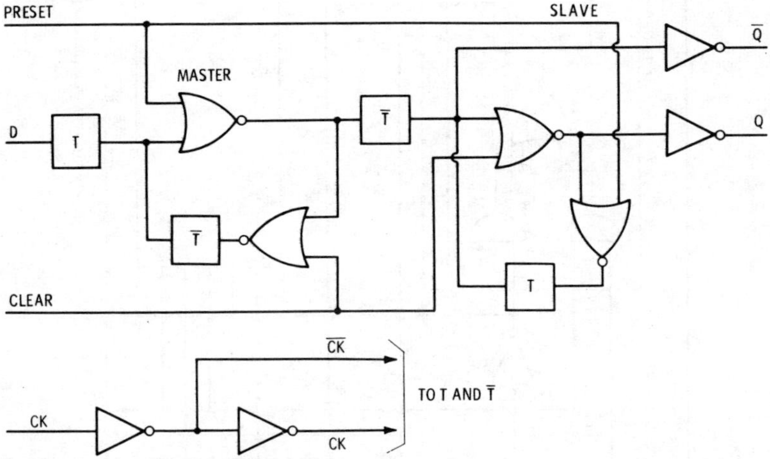

Fig. 3-14. Type 4013, D flip-flop.

Three-State Flip-Flop

Fig. 3-15 shows one of four flip-flops in a type 54C173/74C173 IC. The four outputs can be brought to the high-impedance state by taking either of the two Output Disable signals to logic 1. The circuit also has an Input Disable. When either Input Disable is at logic 0, G1 is 1 and G2 is free to follow input A; the flip-flop sets to A input on a clock transition. When an Input Disable is 1, G1 is 0 and G2 is also 0, regardless of input A. Gate G3, however, is

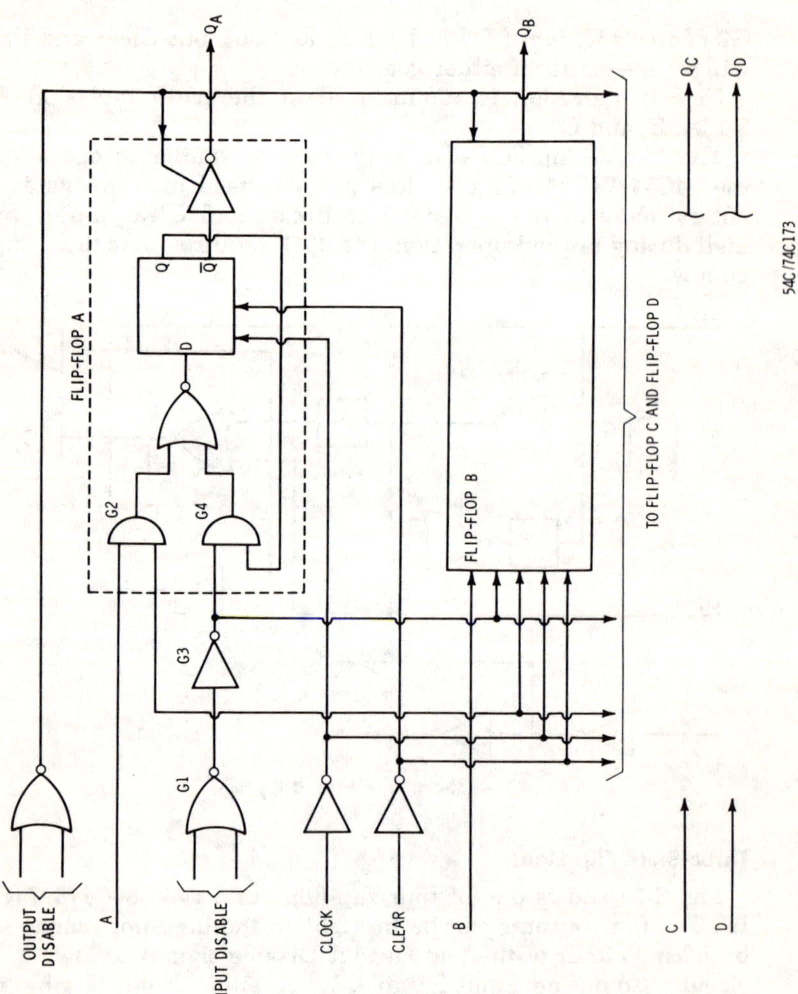

Fig. 3-15. Quad-D flip-flop with three-state output.

1, and this frees G4 to follow Q. Thus, the circuit latches into the Q state whether A is 1 or 0. Clock transitions thereafter cannot affect the state of the D flop.

JK Flip-Flop

The logic diagram of a JK flip-flop, type 4027, is shown in Fig. 3-16A. The transmission gates are controlled by the clock as indicated in Fig. 3-16B. With the clock low, the circuit looks as shown in Fig. 3-17. Gates G6 and G7 are locked in one state or the other, giving Q and $\overline{Q}$. The normal state for CR and P is a low, and, if either goes high, G6 and G7 follow and drive G8 and G9. With CR and P low, G4 follows G3.

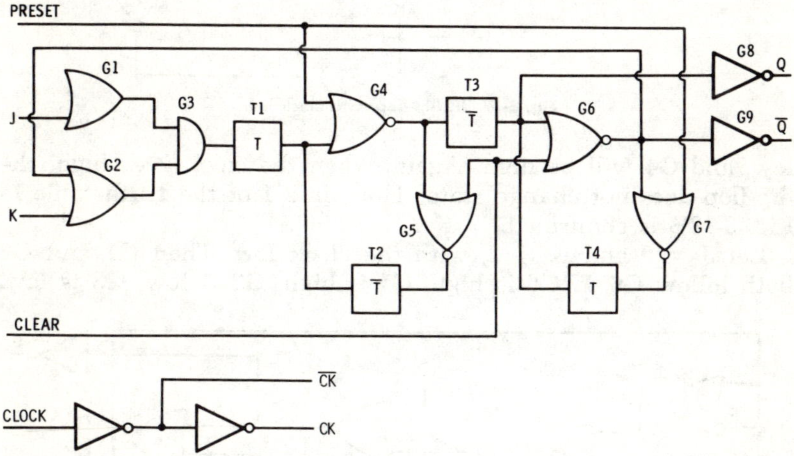

(A) Diagram.

(B) Transmission gates.

CLOCK	T	$\overline{T}$
L	LOW Z	HIGH Z
H	HIGH Z	LOW Z

Fig. 3-16. Logic diagram of type 4027 JK flip-flop.

When the clock is high, the circuit is as shown in Fig. 3-18. Gates G4 and G5 are locked and control Q, and also drive G6 to control $\overline{Q}$. Points CR and P, if high, determine Q and $\overline{Q}$.

With CR and P low, let J = K = 0. In the clock-low state, G2 will be high and G3 will be free to follow G1; G1 in turn will follow G6. Thus, if G6 is high, G1 will be high, G3 will be high, and G4 will be low. When the clock goes high, the low on G4 does not change the state of G6 and the flip-flop does not change state. If G6 was low before the clock, G1 will be low, G3 will be

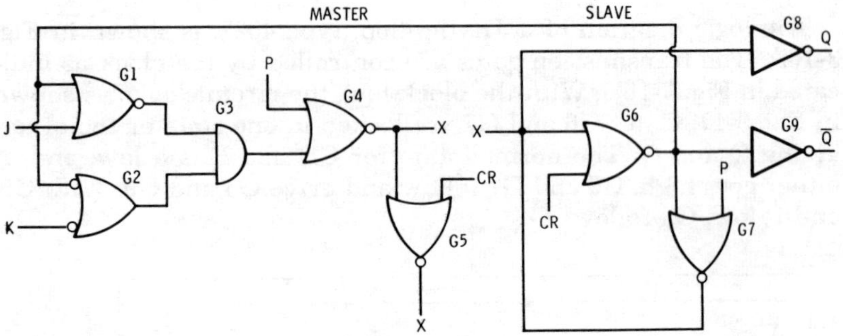

Fig. 3-17. JK flip-flop with clock low.

low, and G4 will be high. Again, when the clock goes high, the flip-flop does not change state. Thus, line 1 of the truth table in Fig. 3-18B is confirmed.

Let J = 0 and K = 1, with the clock low. Then G1 and G2 both follow G6. If G6 is high, G1 is high, G2 is low, G3 is low,

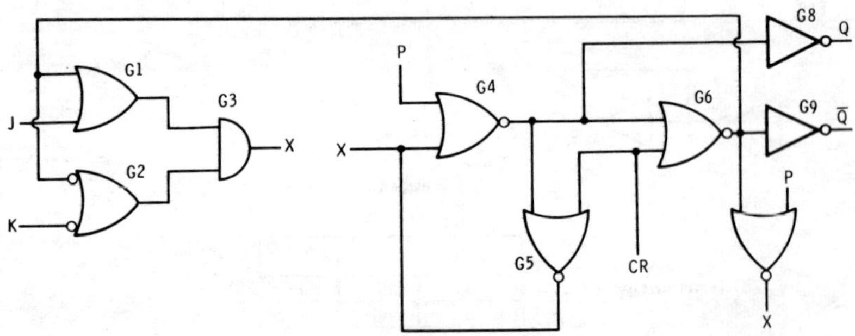

(A) Diagram.

t_n		t_{n+1}	
J	K	Q	$\overline{Q}$
0	0	0	$\overline{Q}$
0	1	0	1
1	0	1	0
1	1	$\overline{Q}$	Q

NO CHANGE (line 1); TOGGLES (line 4)

(B) Truth table.

PRESET	CLEAR	Q	$\overline{Q}$
0	0	Q	$\overline{Q}$
0	1	0	1
1	0	1	0
1	1	1	1

SEE (B) (line 1); NOT ALLOWED (line 4)

(C) Preset and Clear.

Fig. 3-18. JK flip-flop with clock high.

and G4 is high. When the clock goes high, G4 causes G6 to go low and gives Q = 0, $\overline{Q}$ = 1. If G6 was low before the clock, G1 is low, G2 is high, G3 is low, and G4 is high, giving the same result. This is line 2 of the truth table.

For J = 1 and K = 0, with the clock low, G2 is high and G1 is high, regardless of G6. Then G3 is high and G4 is low. When the clock goes high, G6 is driven high, giving Q = 0, $\overline{Q}$ = 1. This is line 3 of the truth table.

For J = K = 1, with the clock low, G1 is high and G2 follows G6. If G6 is low, G2 will be high, G3 will be high, and G4 will be low. When the clock goes high, G6 is driven high. If G6 is high before the clock, G2 will be low, G3 will be low, and G4 will be high. On the clock, G6 is driven low. Thus, the flip-flop changes state, or toggles, on each clock signal. Preset and Clear operations are indicated in Fig. 3-18C.

The circuit of a JK flip-flop in the 54C/74C family (type 54C/74C76) is shown in Fig. 3-19. The circuit is similar to that of the 4027 but has some slight differences. The most important difference is that the flip-flop changes state on the negative-going edge of the clock signal, not the positive-going edge. This is the same operation as the TTL equivalent. In addition, Preset and Clear are performed by low logic levels rather than high.

SHIFT REGISTERS IN CMOS

Shift registers come in many varieties, but all consist essentially of a long chain of storage circuits. In CMOS, shift registers are available with capacities from 4 bits and up. (Very long registers are available in PMOS, both static and dynamic, and will be discussed later.)

An 8-bit shift register will store 8 bits of information, or data. The register may allow data to enter serially, or in parallel, or both, and the outputs may also deliver the data serially, in parallel, or both. Internally, the register may allow the data to be shifted left, right, or both left and right.

Fig. 3-20 shows the block diagram of type 4021, an 8-bit shift register that can be loaded serially at the Serial input and in parallel at inputs A through H. A serial output is available at the 6th, 7th, and 8th bits, allowing the circuit to be used as a 6-, 7-, and 8-bit register. It can also be used as an 8-bit register with 3 bits available in parallel. The input labeled P/$\overline{S}$ puts the register into the parallel load mode when it is high, and into the serial load mode when it is low.

Each bit is stored in a D-type flip-flop, using the circuit shown in Fig. 3-21. When the clock is low, T5 is a high Z and T6 is a low

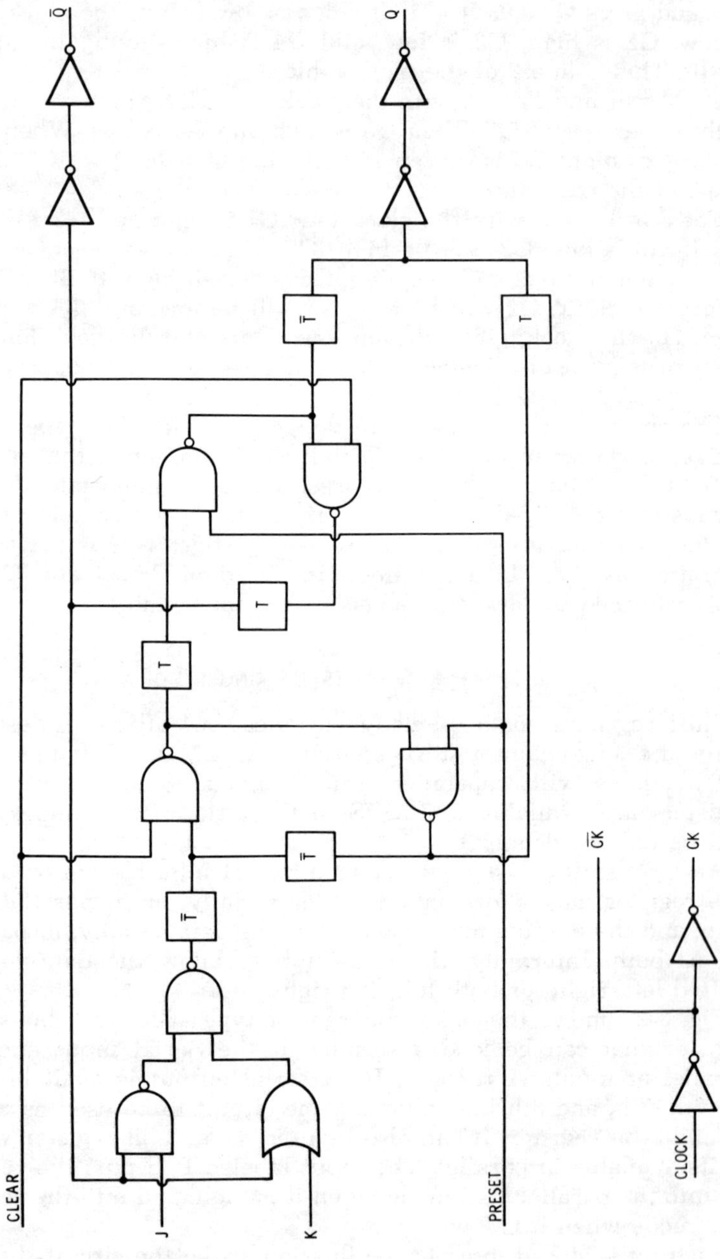

Fig. 3-19. Type 54C76/74C76 JK flip-flop.

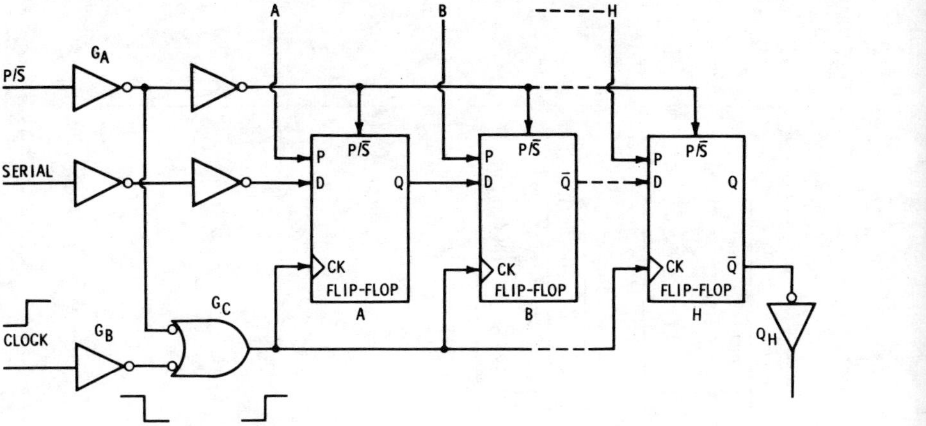

Fig. 3-20. Logic diagram of CMOS shift register, type 4021.

Z, locking G3 and G4 in one state or the other. At the same time, T2 is a low Z, and input D controls the state of G1. When the clock is high, G3 is controlled by G1. Since the clock is high, T2 is a high Z, and therefore the input to G1 is either from P, via T1, or from G2, via T4 and T3. If the input $P/\overline{S}$ is high, then T1 is a low Z and T3 is a high Z; input P controls G1, which in turn controls G3. Thus, data on the parallel inputs are fed into the register.

If $P/\overline{S}$ is low, T1 is a high Z, and T3 is a low Z; when the clock is high, T4 is also low, and therefore G1 and G2 are latched into the state previously determined by input D.

Looking at the overall block diagram, Fig. 3-20, when $P/\overline{S}$ is high, G_A is low and thus the output of G_C is held high. Thus, the clock input to each flip-flop is high, regardless of the actual state of the system clock. Parallel data have therefore been loaded into the register. When $P/\overline{S}$ is low, G_A is high and G_C is free to follow G_B, which follows the clock. Thus, the data at the serial input are entered into flip-flop A when the system clock goes from low to high. At the same time, the data in flip-flop A are transferred to flip-flop B, and so forth throughout the register.

Left-Shift, Right-Shift Register

Type 54S/74C95 is a CMOS version of TTL type 54/7495, and the same truth table applies to both, but the operating speeds, power dissipation, etc., are different.

When mode control input MC is low (Fig. 3-22), data are entered at the serial input and shifted right on the negative-going edge of the Clock 1 input. All the other inputs are inhibited, but the output is available in parallel or serially. When MC is high,

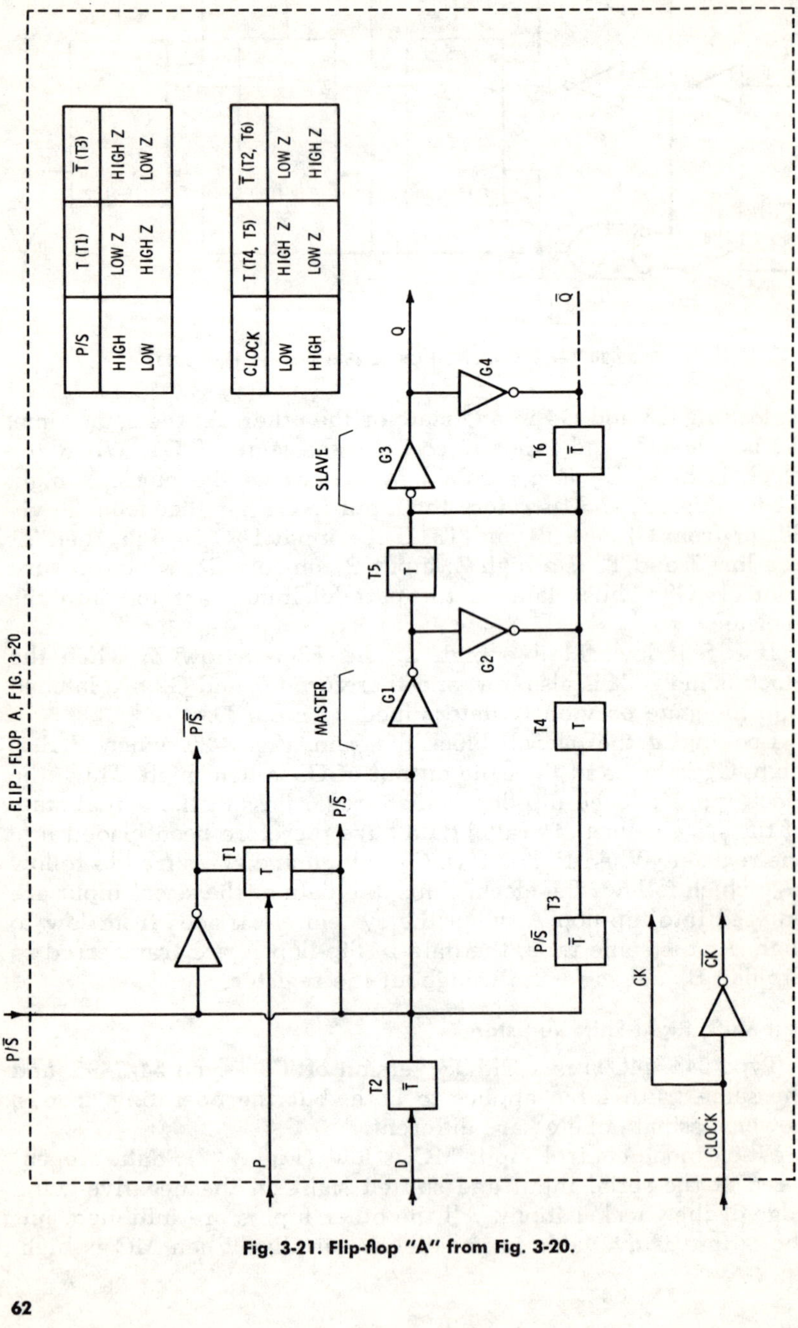

P/S	T (T1)	$\overline{T}$ (T3)
HIGH	LOW Z	HIGH Z
LOW	HIGH Z	LOW Z

CLOCK	T (T4, T5)	$\overline{T}$ (T2, T6)
LOW	HIGH Z	LOW Z
HIGH	LOW Z	HIGH Z

Fig. 3-21. Flip-flop "A" from Fig. 3-20.

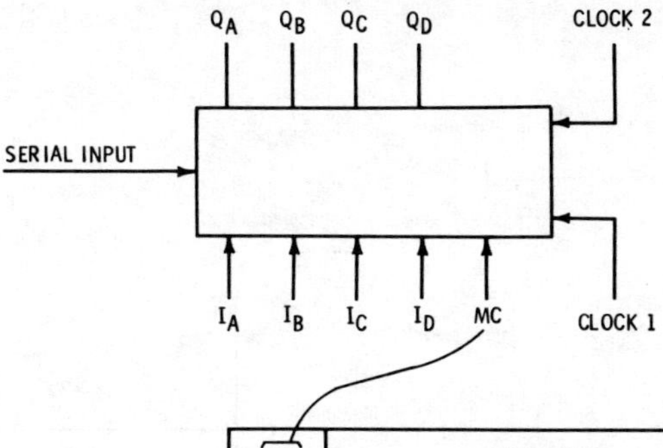

$\overline{MC}$	OPERATION
0	DATA ENTERS AT SERIAL INPUT. SHIFTS RIGHT ON NEGATIVE-GOING EDGE OF CLOCK 1. OTHER INPUTS ARE INHIBITED.
1	PARALLEL DATA IN AND PARALLEL DATA OUT ON NEGATIVE EDGE OF CLOCK 2. OTHER INPUTS ARE INHIBITED.

Fig. 3-22. CMOS left-shift, right-shift register.

the parallel inputs and Clock 2 are enabled, while the serial input and Clock 1 are disabled. This allows parallel in and parallel out on the negative edge of Clock 2.

For left-shift operation, the parallel outputs are connected back to inputs as shown in Fig. 3-23, and the output is taken serially at Q_A.

Other shift registers operate similarly to the preceding but may include circuits that allow clearing the register to all 0s, or have JK-type inputs where the state of the JK inputs determines whether a 1 or a 0 is entered into the register—that is, the data can be entered in inverted form.

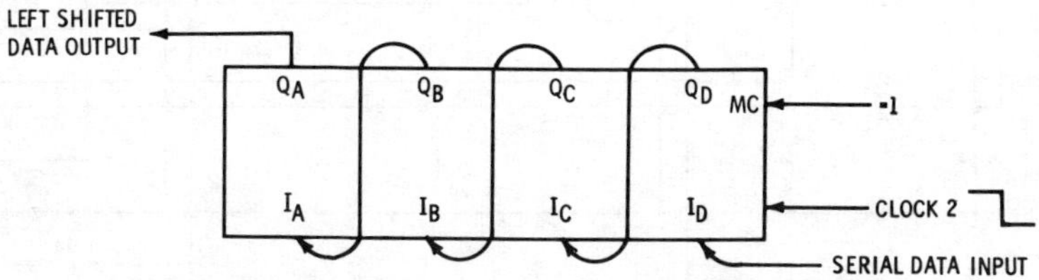

Fig. 3-23. Connections for left shift on a CMOS register.

CMOS COUNTERS

Counters in CMOS have maximum operating speeds only up to about 5 MHz for $V_{DD} = +10$ V, and lower speeds at $V_{DD} = +5$ V, but the functional complexity can be high.

Type 4029, shown in Fig. 3-24, will count in the binary or decade mode, will count up or down, and is presettable to any de-

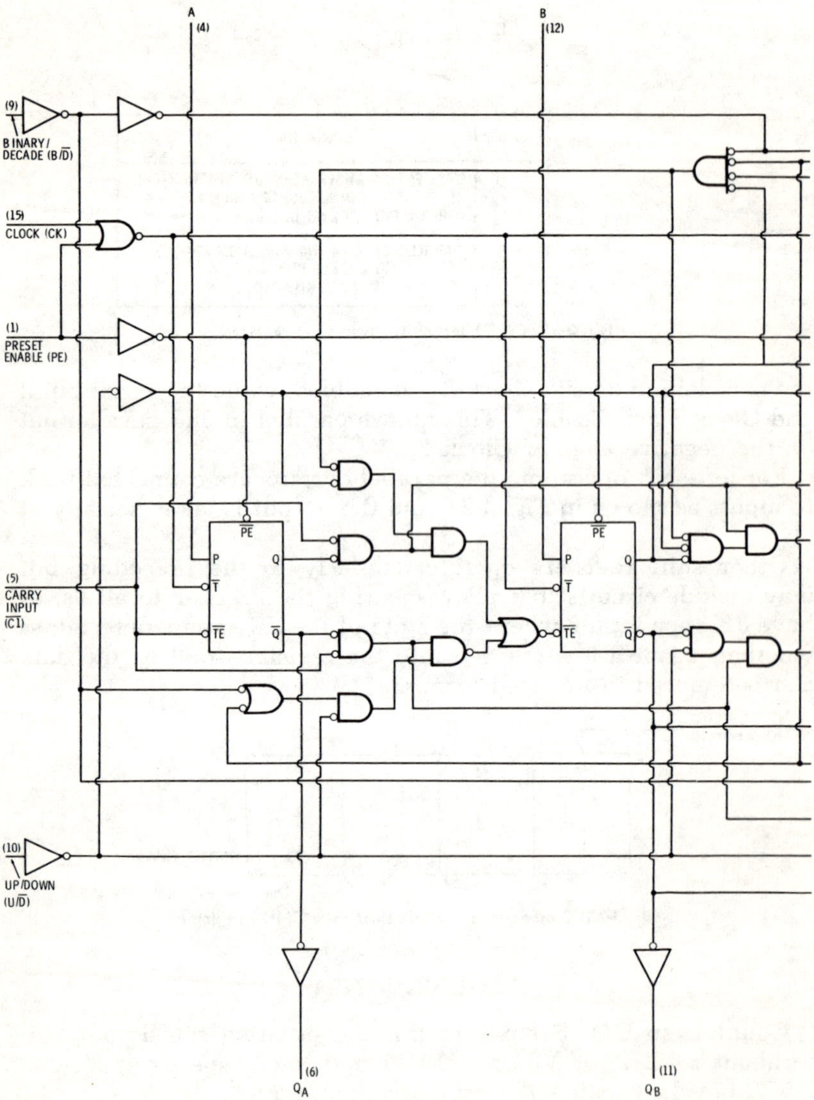

Fig. 3-24. Logic diagram for a CMOS up/down,

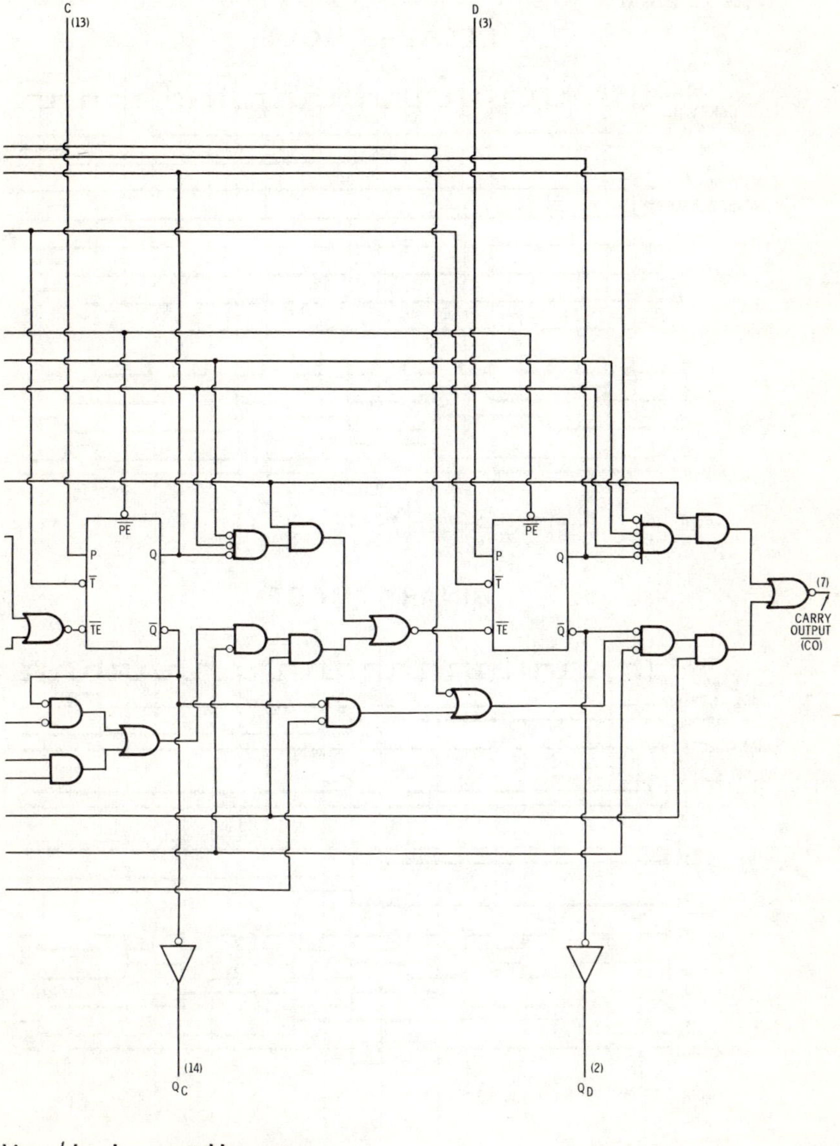

binary/decade, presettable counter.

sired count. The timing diagram is shown in Fig. 3-25. Because the circuit is complex, it is shown in block form in Fig. 3-26. The control block determines how the four flip-flops of the circuit will respond. The action of this block is detailed in the operating-mode table of Fig. 3-27. When the B/D input is high, the circuit counts in binary; when it is low, it counts in decimal or decade.

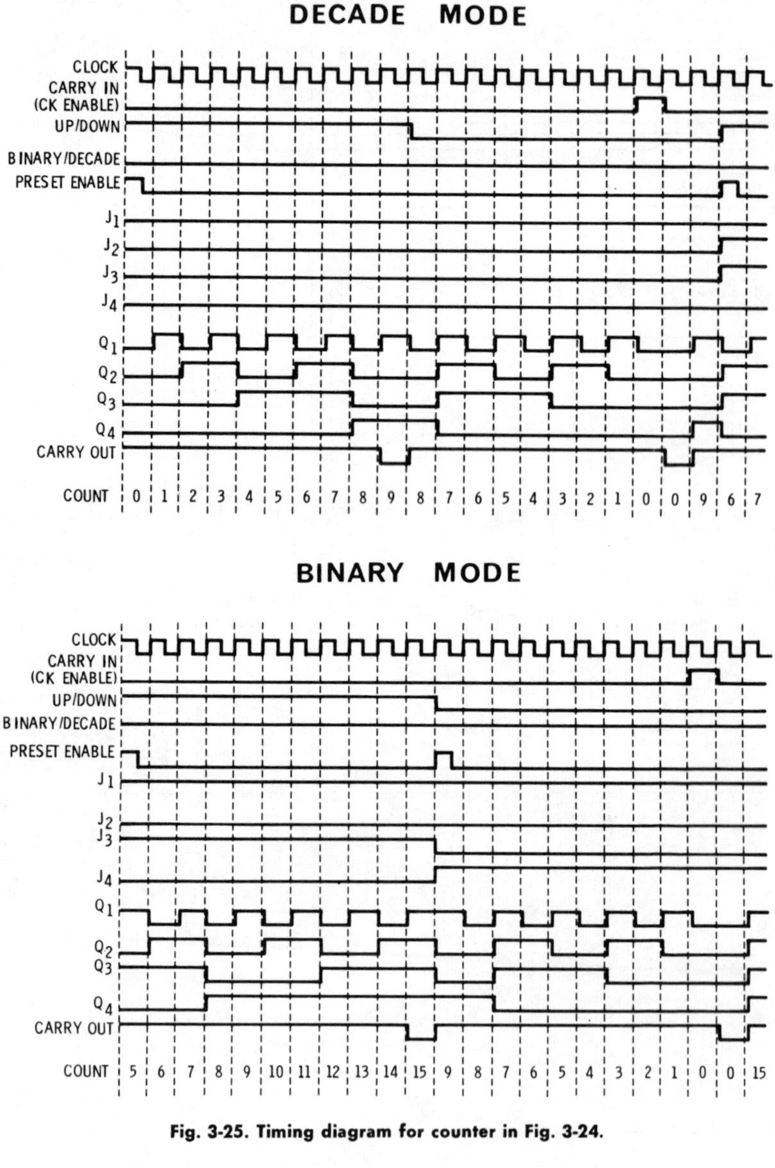

Fig. 3-25. Timing diagram for counter in Fig. 3-24.

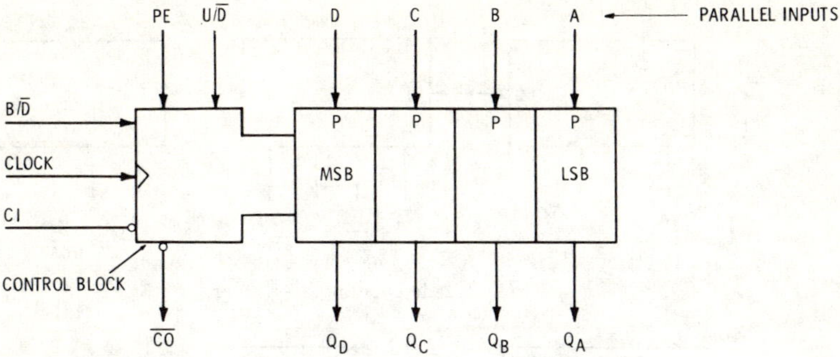

Fig. 3-26. Block diagram of counter in Fig. 3-24.

When the $U/\overline{D}$ input is high, the flip-flops count up; when it is low, they count down. When PE is high, the flip-flops will be set individually to the levels of the four inputs A, B, C, and D; when PE is low, the circuit is allowed to count in the mode determined by the other inputs. Input $\overline{CI}$ is a carry input of less significant counts from a preceding stage. If $\overline{CI}$ is low, the flip-flop will count normally on the negative-going edges of the clock. If $\overline{CI}$ is high, counting is inhibited.

When the counters are connected for decade counting as shown in Fig. 3-28, C1, when full, makes carry output CO low; thus, every time C1 is full, it allows one count to be entered in C2. Counter C2, when full, in turn allows one count into C3, and thus the count builds up. In Fig. 3-28A, the clock is applied to all the ICs in parallel, which provides the fastest operation. Ripple clocking can also be used, as in Fig. 3-28B; this allows the clock to have longer rise and fall times.

Each flip-flop within the counter operates as shown in Fig. 3-29. When $\overline{PE}$ is low, the flip-flop follows input P, and thus the stage is preset to the level of the parallel input (A, B, C, or D). Note

CONTROL INPUT		OPERATING MODE
$B/\overline{D}$	H	BINARY COUNTER
	L	DECADE COUNTER
$U/\overline{D}$	H	COUNT UP
	L	COUNT DOWN
PE	H	PARALLEL LOAD
	L	ENABLE COUNTER
$\overline{CI}$	H	INHIBIT COUNTER
	L	ENABLE COUNTER

Fig. 3-27. Mode table.

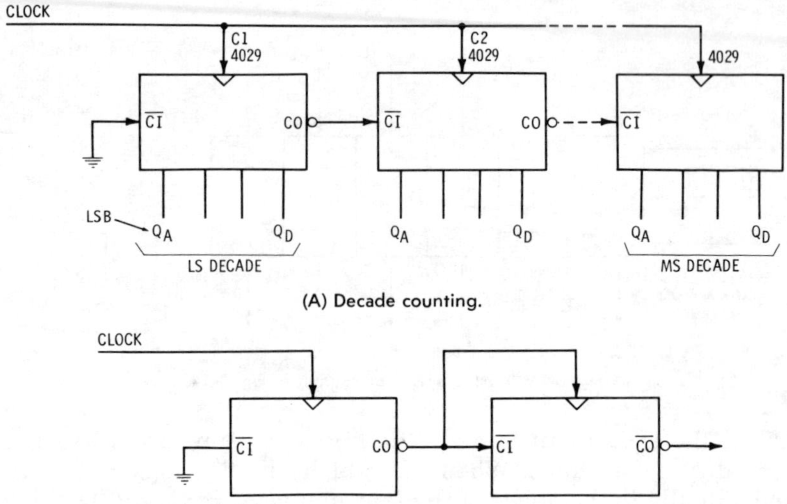

(A) Decade counting.

(B) Ripple counting.

Fig. 3-28. Connections for decade counting and ripple counting.

from Fig. 3-24 that when $\overline{PE}$ is low, it also drives the $\overline{T}$ input low, which the flip-flop requires if P is to control the output. When $\overline{PE}$ is high, the flip-flop responds to input $\overline{TE}$ on a negative-going transition of $\overline{T}$; output Q does not change if $\overline{TE}$ is high but does change if $\overline{TE}$ is low. When input $\overline{T}$ is high, the flip-flop is stable and does not respond to any input.

The external gates around the flip-flops then act to steer the signals in such a way that counter operation specified by the control inputs is obtained.

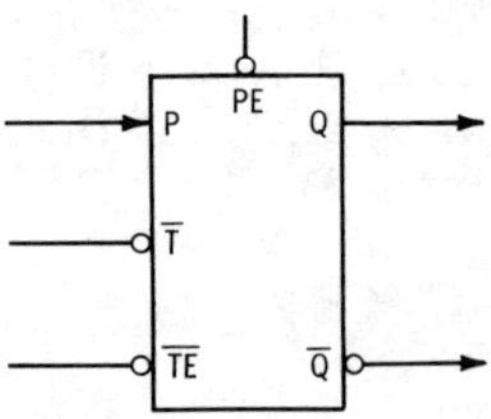

$\overline{PE}$	$\overline{TE}$	$\overline{T}$	P	Q	$\overline{Q}$
L	X	L	H	H	L
L	X	L	L	L	H
H	H	↓	X	Q_n	$\overline{Q}_n$
H	L	↓	X	$\overline{Q}_n$	Q_n
X	X	H	X	Q_0	$\overline{Q}_0$

Fig. 3-29. Counter stage and truth table.

Johnson Counter

A counter somewhat simpler than the foregoing one is shown in Fig. 3-30A. Here a 5-stage Johnson counter, type 4017, is used to provide ten outputs, which go high, one after the other. When Clear goes high, all the flip-flops are set to Q = 0. This is decoded by the ten output gates. For E = A = 0, only the Y0 gate is satisfied and therefore Y0 goes high.

With the Clock Enable low, positive-going clock transitions are fed to the flip-flops. From the feedbacks used, the flip-flops follow the sequence shown in the truth table of Fig. 3-30B. On the first clock transition, flip-flop A goes to Q = 1 and all the rest stay low. This condition of A = 1 and B = 0 satisfies gate Y1, which goes high, while Y0 goes low. Succeeding clock signals fed into the circuit change the flip-flops as indicated, and each state satisfies only one of the ten output gates. Circuit operation is seen to be similar to a stepping switch. Note that the Carry output, CO, goes high on count 0 and stays there until count 5 is reached.

Other Counters

In the 54C/74C family, both binary and decade counters are available to do the same job as the 4029, but they are not available as one device which can be used in so many ways. Such a universal counter can be made in 54C/74C if a manufacturer decides to do so. Some CMOS counters also have 3-state outputs: type 75C/85C55 is a presettable decade counter with this type of output.

CMOS DECODERS, MULTIPLEXERS, AND DEMULTIPLEXERS

Decoders and multiplexers operate similarly. The main difference is that a decoder operates on a set of inputs to drive a specific output or outputs high or low, while a multiplexer operates on a set of inputs to select a specific output, which will then be driven high or low, depending on the state of a data input. In a decoder, the set of inputs is called the "code," while in the multiplexer it is called the "address."

Fig. 3-31 shows the logic of type 54C/74C42, which is a decoder that converts binary-coded-decimal to decimal. Circuit operation is very straightforward. When inputs A, B, C, and D are all 0, output 0 is selected and goes low, while all the other outputs stay high. With all inputs 0, G2 is the only gate with both its inputs 0, and thus it goes high and the output goes low. For other input codes, the circuit responds as shown in the truth table.

Type 4028 (not shown) is similar to the 54C/74C42 except that it has extra inverters on its outputs. As a result, the decoded output goes high while all the others stay low.

Type 54C/74C151, shown in Fig. 3-32, is an 8-channel multiplexer. The 3-bit data-select input signal allows one and only one of the data inputs to propagate through the circuit to the outputs. When the Strobe input is 0, G1 and G2 are free to follow input C, and the logic provides the selected data. When Strobe is 1, the

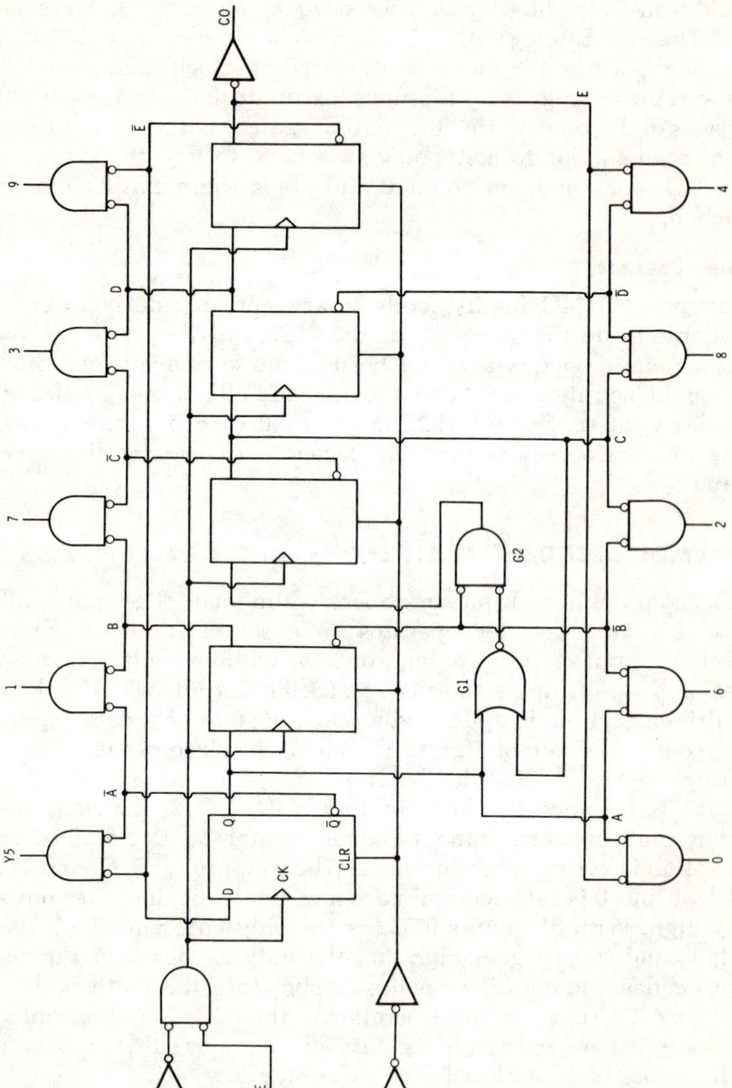

(A) Logic diagram.

Fig. 3-30. Johnson

circuit locks out the data inputs, and the W output goes low and Y goes high. With Strobe 0 and with inputs A, B, C, and D all 0, the circuit allows the data on input D0 to appear at output W, and $\overline{D0}$ to appear at Y. As the input address changes, the outputs follow the selected data input.

A somewhat different scheme is used with type 4512, which is an 8-channel data selector, or multiplexer, with a 3-state output. When the transmission gate (Fig. 3-33) has a 0 at the open-triangle input and a 1 at the other control input, the gate presents a low Z at the signal terminal; for the opposite control input, the impedance is high. When all the address inputs S0, S1, and S2 are low, the three transmission gates in the path from D0 to the output are all in the low-Z state, while all the other D signals are blocked by one or more high-Z gates. If the Enable ($\overline{E}$) input is low, G1 is free to respond to the selected D input. If the output control (OC) signal is low, output Y goes to the same level as the selected input. If the output control signal is high, Y goes to the high-Z state.

Once the data input signal has passed through its associated input inverter, it then must pass through three transmission gates in series before it goes through another stage of amplification in G1. Within an IC, the signal lost in passing through these gates is of little importance. Note that the transmission gates do not delay the signal.

A 4-line to 16-line decoder/demultiplexer, type 54C/74C154, is shown in Fig. 3-34. When control inputs G1 and G2 are both 0,

	CK ↑	A $\overline{A}$	B $\overline{B}$	C $\overline{C}$	D $\overline{D}$	E $\overline{E}$	HIGH Y	CO
(AFTER CLEAR) →	0	0 1	0 1	0 1	0 1	0 1	Y0	H
	1	1 0	0 1	0 1	0 1	0 1	Y1	H
	2	1 0	1 0	0 1	0 1	0 1	Y2	H
	3	1 0	1 0	1 0	0 1	0 1	Y3	H
	4	1 0	1 0	1 0	1 0	0 1	Y4	H
	5	1 0	1 0	1 0	1 0	1 0	Y5	L
	6	0 1	1 0	1 0	1 0	1 0	Y6	L
	7	0 1	0 1	1 0	1 0	1 0	Y7	L
	8	0 1	0 1	0 1	1 0	1 0	Y8	L
	9	0 1	0 1	0 1	0 1	1 0	Y9	L

(B) Truth table.

counter, type 4017.

the four gates, GX, are free to follow their associated inputs, C or D. The 16 outputs will then be selected in accordance with the input code. If A, B, C, and D are all 0, output 0 is selected and goes to 0, while the other outputs stay high. With output 0 selected, let input G2 go high. This, in turn, causes output 0 to go high. Thus, the selected output will follow the signal on G2. Similarly, if data are applied to G1 instead of G2, the result is the same. Either G1 or G2 input, therefore, can be used as a data input, while the other is used as an enable input. When the enable input is low, the selected output follows the data input.

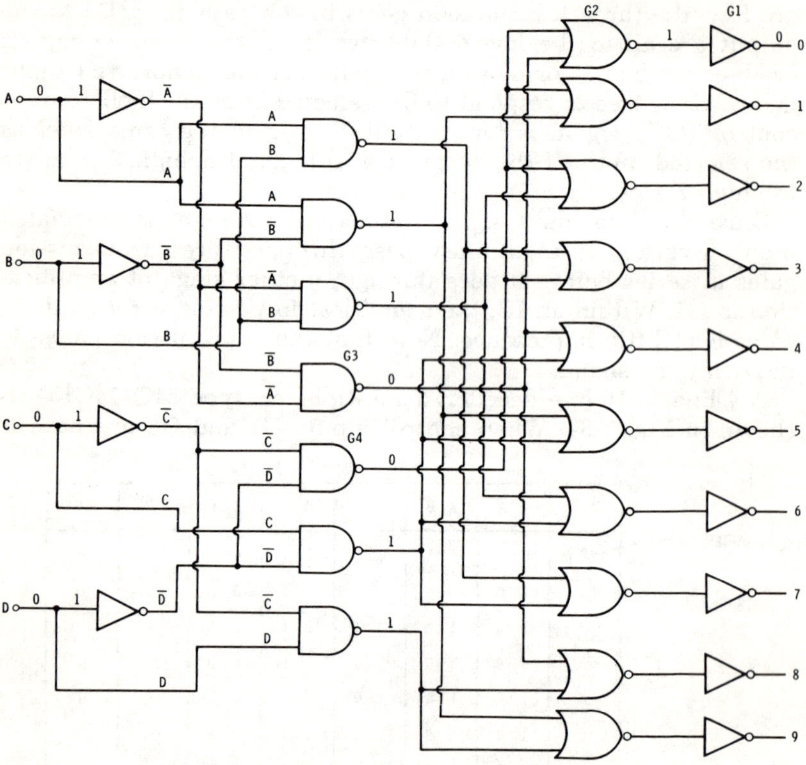

Fig. 3-31. Binary-coded-decimal to decimal decoder.

CMOS Decoder With Bipolar Outputs

Because of their relatively high output impedance, CMOS circuits are not very good at driving loads requiring more than a few hundred microamperes. Yet an important class of loads are the lamps used in displays, and many of these require more current than a CMOS output stage can deliver.

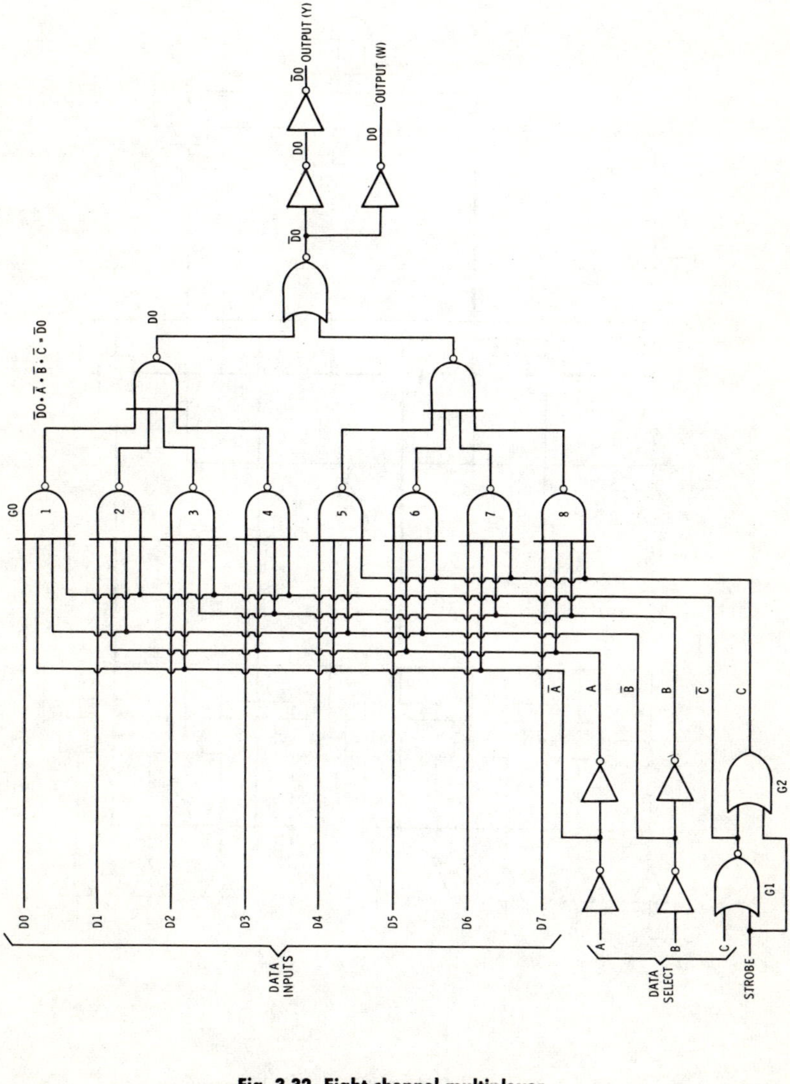

Fig. 3-32. Eight-channel multiplexer.

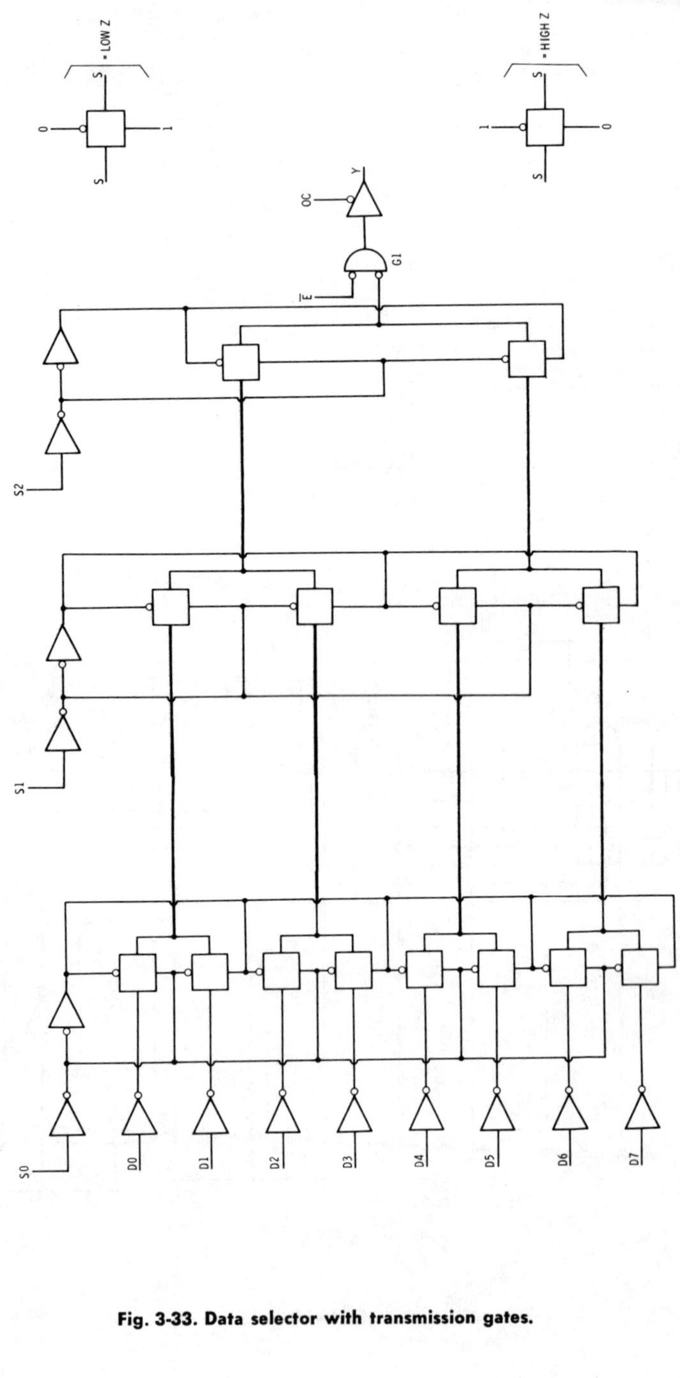

Fig. 3-33. Data selector with transmission gates.

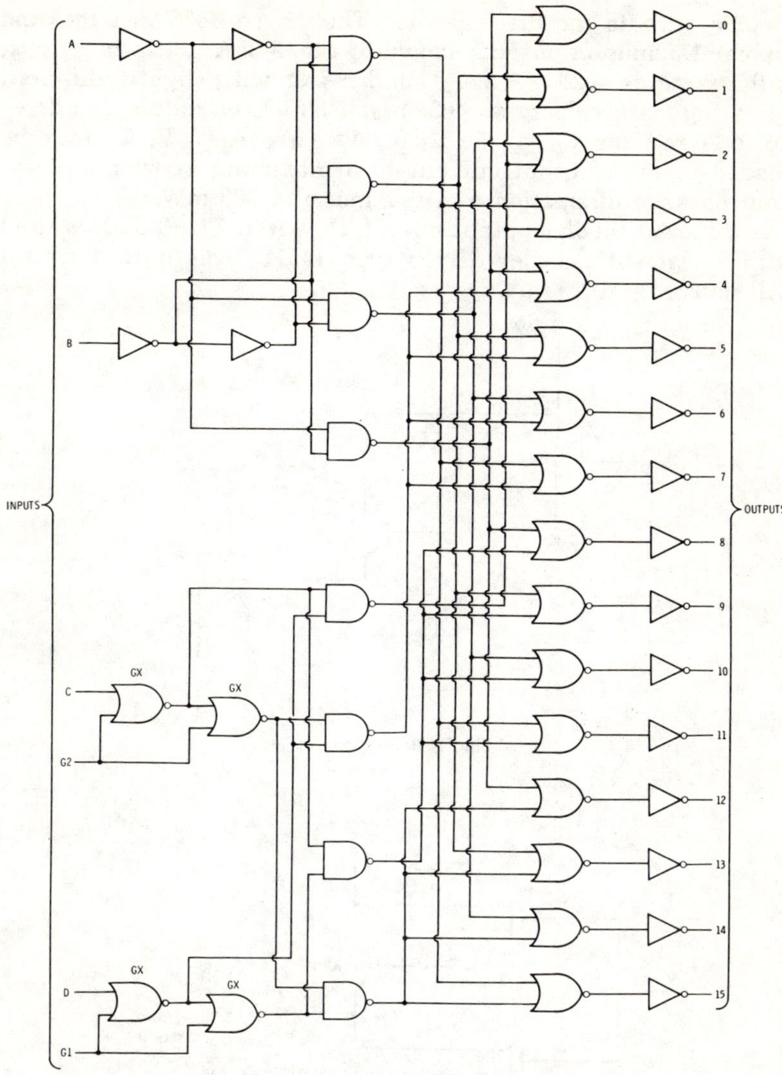

Fig. 3-34. Four-line to 16-line decoder/multiplexer.

Type 4026, for example, is a device that acts as a decade counter, decoder, and driver for low-power 7-segment displays. It is similar functionally to TTL type 54/74143 (and similar TTL ICs), but while the TTL IC can sink 15 mA for each segment, the CMOS IC can sink only about 0.6 mA. To provide more drive capability, certain ICs combine CMOS logic for complexity with bipolar outputs for drive power. Thus, type 4426 has the npn bipolar Darlington outputs shown in Fig. 3-35A, but otherwise it is the same as 4026. (A very similar pair with slightly different operating features is types 4033 and 4433.) Each output can drive up to 20 mA for $V_{DD} = +5$ V; for V_{DD} above +5 V, Rx may be needed to limit output current to a maximum of 60 mA or to limit package dissipation to a maximum of 300 mW.

A different bipolar output circuit, shown in Fig. 3-35B, is used with 7-segment decoder/driver type 14511. This output circuit will source up to 25 mA.

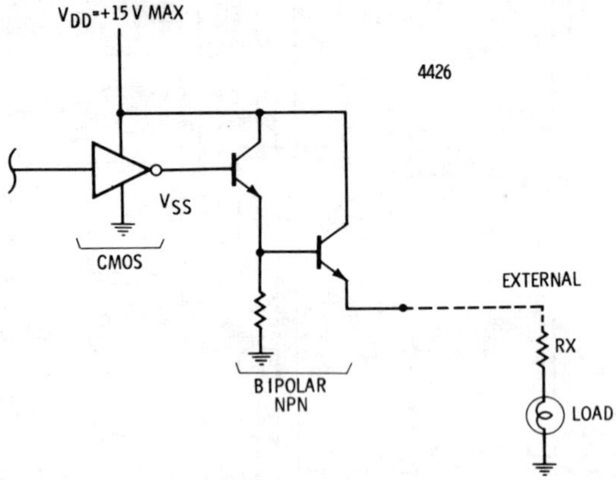

(A) Bipolar Darlington output.

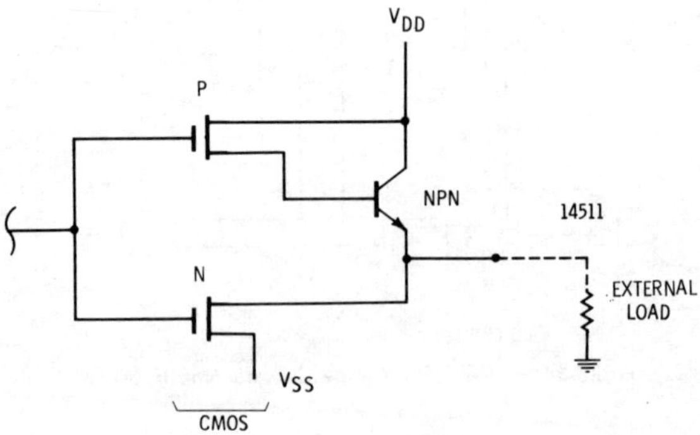

(B) Bipolar transistor output.

Fig. 3-35. Bipolar outputs for CMOS logic.

The combination of bipolar and MOS transistors in the same IC can be expected whenever the operating advantage thus obtained is worth the extra processing required during manufacture. Both technologies use pn junctions as a starting point and thus are compatible.

CMOS DUPLICATES OF TTL CIRCUITS

The equivalence between 54C/74C CMOS ICs and ICs in the 54/74 family of TTL is readily apparent because the same part numbers are used in both families. All devices with the same numbers have the same pinouts and perform identical functions. However, you cannot unplug a 54/7400 and plug in a 54C/74C00 and expect the thing to work. While the voltage levels are compatible, the current levels are not.

One 54/7400 can drive 10 TTL inputs, but one 54C/74C00 cannot even drive one regular TTL input. The best it can do is drive two low-power TTL loads. But if all the devices in a 54/74 TTL design are replaced by 54C/74C devices—assuming they exist—then the logic part of the circuit will function correctly but the input and output connections to the outside world may have to be modified. Also, of course, the CMOS version will run much slower than the TTL version.

An equivalence also exists between 4000 and 14500 series CMOS and other logic circuits. Devices in 54C/74C act on the same edge of the signal, positive or negative, as the same device in 54/74. Equivalent devices in 4000 and 14500 may or may not act on the same edge.

In all the CMOS families, circuits are available that perform most of the same functions as devices in various TTL families. You can expect that most or all of the functions available in TTL will also be made in 54C/74C, in 4000 and 14500 series devices, and possibly in other CMOS families. These functions include a large variety of gates, inverters, buffers, flip-flops, data selectors, shift registers, arithmetic circuits, comparators, multiplexers, priority encoders and decoders, latches, multivibrators, RAMs, ROMs, PROMs, CAMs, etc. Truth tables for the CMOS versions of such circuits will be identical in many cases to the TTL versions, although the internal logic may differ.

CMOS ICs can pack more complexity into a chip than TTL can, and while operations are approximately the same, CMOS ICs may include extra operating features such as 3-state outputs. CMOS ICs tend more and more to include data storage in such devices as shift registers and the various memories. Type SCL5533, for example, is a 64-bit CAM by Solid State Scientific, Inc., which operates similarly to type 93402, a 16-bit TTL CAM.

4

PMOS Technology

CHARACTERISTICS OF PMOS ICS

The first type of MOS to become widely available in IC form used p-channel enhancement-mode transistors. ICs using PMOS technology have a number of characteristics in common, but they also have many differences. A relatively large number of families exist, some using high-threshold transistors and some low-threshold, some using metal gates and some using silicon gates, some using ion implantation and some not. The most significant characteristics of PMOS ICs are summarized as follows.

1. The ICs operate with negative voltages of as much as −30 V or so. In the operation of any specific PMOS transistor, potential difference is the main consideration, not absolute voltage levels; if one level is +100 V and the other is +70 V, a satisfactory situation exists and proper operation can occur. A problem arises when PMOS ICs are working with bipolar circuits such as TTL, which operate with a 5-volt differential. In this case it is often desirable to operate the TTL circuit with +5 V and 0 V, and the PMOS circuit with +5 V and −25 V. Many of the newer PMOS ICs work with +5-V and −12-V power supplies, sometimes with a connection to the 0-V point and sometimes without. While some commonality of power supplies is thus obtained, it does not mean that logic levels are also compatible.
2. Most, but not all, PMOS ICs require more than one power supply. If V_{SS} is taken as the reference voltage (typically, 0 V), which is usually but not always true, a separate supply

is required for V_{DD} and V_{GG}. Voltages V_{DD} and V_{GG} usually cannot be obtained from a simple resistance divider, since each must be independently regulated, usually to ±5%.

3. The operating speed of PMOS ICs is slow compared with bipolar devices. The access time of many PMOS memory ICs is around 500 ns, although some recent devices have times of about 100 ns. The access time of a typical bipolar TTL RAM such as type 54/7484 (16 bits) is about 20 ns. A JK flip-flop in standard TTL can toggle at 40 MHz, whereas a typical JK flip-flop in PMOS is guaranteed to toggle only up to 250 kHz—a difference in speed of 160 to 1.
4. Because PMOS transistors can be made very small, a great many can be put down on one chip. While these transistors can be connected to each other fairly easily on the chip, the connection between the chip and the outside world takes up a lot of space. As a result, the number of connections to the chip is kept as small as possible, while the number of transistors on the chip is made as large as possible. In the process of putting many transistors in a small area, regular arrays of rows and columns simplify the layout and mask-registration problems, therefore increasing yield. The circuits that best meet these requirements are shift registers and memories, and a large number of PMOS ICs are memory devices.
5. A significant number of PMOS shift registers and memories are dynamic, which means they require clock signals regularly. Sometimes a single clock signal at TTL logic levels is specified, but many circuits require two interrelated clock signals at PMOS voltage levels. Still other circuits require three clock signals, and a few require four. Dynamic circuits pack a lot of bits in a small area and do not use much power, but they require external support circuits and a certain amout of babying.
6. Input circuits of PMOS ICs (and other MOS ICs) are largely capacitive. This means that the circuits tend to draw spikes of current to charge the capacitive inputs or to discharge already charged capacitances. Unless the current spikes are effectively controlled, they can be coupled into nearby or otherwise-connected circuits where they can cause improper operation. Since the circuits contain very little resistance to dissipate energy, ringing can easily result. Ringing can cause an individual circuit to malfunction and may also be propagated through to affect other circuits.

Since most of the current used is drawn in spikes at clock transitions, the average current is a function of operating frequency. This is an advantage when the operating frequency is low. When PMOS circuits operate near their upper frequency limits, they may use as much average power as an equivalent low-power TTL circuit.

The voltage levels needed at the signal inputs of PMOS are of two types. Logic-level inputs such as addresses, data inputs, chip select, etc., are near the logic levels of bipolar logic, and sometimes bipolar logic levels can be used directly. Sometimes they can be used by adding a pull-up (or pull-down) resistor, and sometimes these are built into PMOS ICs. The second type of inputs are the clocks. Clock input signals can be TTL levels in some cases but in others must be at PMOS voltage levels.

7. The outputs of many PMOS circuits are voltage levels—high or low—as in bipolar logic circuits. Much effort is spent in obtaining outputs that are directly usable with bipolar logic (as well as with PMOS, of course), and sometimes this objective is obtained and sometimes not. The outputs of a significant group of PMOS memories are not voltage levels at all but are either a high current or pulse for an H and a low current or pulse for a L. These signals are not generally compatible with anything and must be amplified and converted to voltage levels before they can be further processed. These outputs are similar to core memory outputs.

   The output impedance of most PMOS ICs is high compared with bipolar devices. Some outputs are barely able to drive one low-power TTL input, which requires only 160 μA; others can drive one regular TTL input (1.6 mA), and a few can drive two regular TTL inputs.

PMOS BASIC CIRCUITS

Since resistors require more room on a chip than a MOS transistor, the common load element in a p-channel enhancement-mode IC is a transistor with its gate permanently connected so as to bring the device into its conducting region. In the basic inverter circuit, this can be done as shown in Figs. 4-1A and B.

In Fig. 4-1A, the gate of Q_L is tied to V_{DD}, while in Fig. 4-1B it is tied to a separate source of negative voltage. In Fig. 4-1A, as V_{IN} is brought toward ground, Q_A turns off and V_O approaches V_{DD}. As V_O comes close to V_{DD}, the gate-to-source voltage differential of Q_{LA} becomes less and less, and the impedance of Q_{LA} increases. If V_O could get close to V_{DD}, Q_{LA} would turn off. This type of operation is satisfactory for some circuits. But if V_O is trying to drive current into a capacitor, the increasing impedance

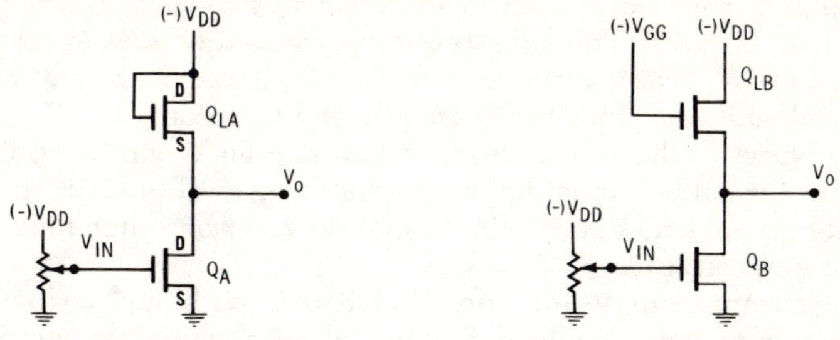

(A) Saturated.

(B) Unsaturated.

Fig. 4-1. Basic PMOS inverter.

of Q_{LA} as V_O approaches V_{DD} causes the charging rate to decrease, and thus slows the circuit.

In the circuit of Fig. 4-1B, the gate of Q_{LB} is supplied by V_{GG}, which is made more negative than V_{DD} by at least the threshold voltage of Q_{LB}. Now when V_O approaches V_{DD}, the impedance of Q_{LB} increases only slightly. The circuit at B is considerably faster than the circuit at A. The circuit at A is said to operate in the saturated mode, and the circuit at B in the unsaturated mode.

Voltage levels for V_{GG} are usually about 5 V more negative than V_{DD}, although some ICs have specifications with V_{GG} up to 10 V more negative than V_{DD}.

When Q_B is on, its impedance must be much lower than the impedance of Q_{LB} so that V_O can swing from near V_{DD} to near ground. To meet this requirement, Q_B is made physically larger. (Similarly for Q_A.)

Not many simple gates have been made in PMOS because the technology is best suited to complex circuits. But type 3102 (Fig. 4-2), made by Fairchild Semiconductor Corp., is a 3-input PMOS

Fig. 4-2. PMOS logic gate.

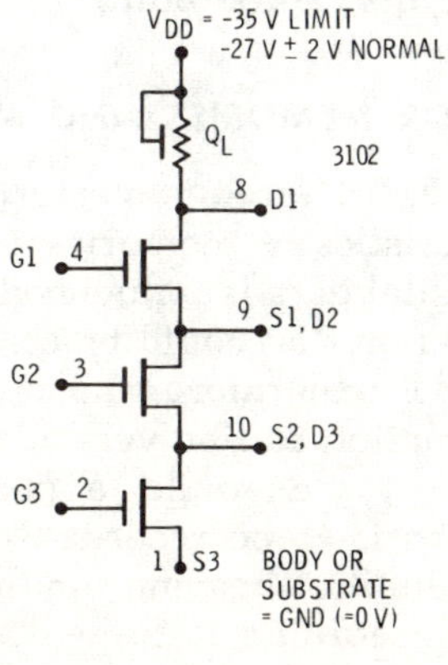

gate that can be connected in several ways to form NORS, NANDS, and other circuits. The substrate or body is the voltage reference point, or 0 V. The source and drain of all the transistors are p material, and therefore these are allowed to go only 0.3 V positive with respect to the substrate. As much as −35 V can be applied to any of the other pins before breakdown will occur, but for normal operation, V_{DD} is −10 to −27 V, and the other pins act as inputs and outputs.

The circuit consists of four transistors, with Q_L permanently connected to act as a load for the other three transistors. The resistance of Q_L is about 140K; if V_{DD} is set at −25 V and pin D1 is connected to ground, about 180 μA will flow through Q_L.

When the circuit is connected as shown in Fig. 4-3A, it acts as a 3-input NAND gate. If all three transistors are on, each will have an impedance of about 500 ohms and the output voltage at pin 8 (that is, Y) is about $[1500/(140{,}000 + 1500)] \times 27 = 0.286$ V. The actual voltage at Y will depend on how much current the circuit is required to sink. If the sink current is 1 mA, the voltage at Y will be about $0.286 + 1.5 = 1.78$ V.

When connected as in Fig. 4-3B, the circuit is a 3-input NOR gate. If any one of the three transistors is on, Y goes low.

How much gate voltage is needed to turn on one of the input transistors? This is given by Fig. 4-3C. When a gate is about 5 V negative with respect to its source, each transistor begins to turn on and has a resistance of about 2K. As the gate-to-source voltage differential is increased, the "on" resistance falls to about 400 ohms. Logic levels for these gates can be from 0 to −2 V for a logic H and from about −9 to −20 for a logic L (for $V_{DD} = -27$ V).

The device can act as an inverter buffer (Fig. 4-3D) or a noninverting buffer (Fig. 4-4E). By combining several devices, other simple logic circuits can be built.

SEMICONDUCTOR MEMORIES AND MEMORY COSTS

Because so many MOS circuits perform memory functions, some overall characteristics of memories need to be considered. The cost of storing digital data is influenced by many factors, and the amount of information that could be converted to digital form and stored is infinite. Furthermore, the number of bits that are needed to store information can be very high for certain types of information. Consider, for example, a performance by a symphony orchestra, which is to be recorded digitally.

The average upper limit of hearing for the human ear is about 17 kHz. If the digital recording is to be able to reproduce these waveforms accurately, the audio signal must be sampled at least twice in each waveform (to prevent a form of error called aliasing or frequency folding), which means 34,000 times per second. If the recording is to have high fidelity, each sample must be as accurate as possible. If each sample is digitized to 10 bits,

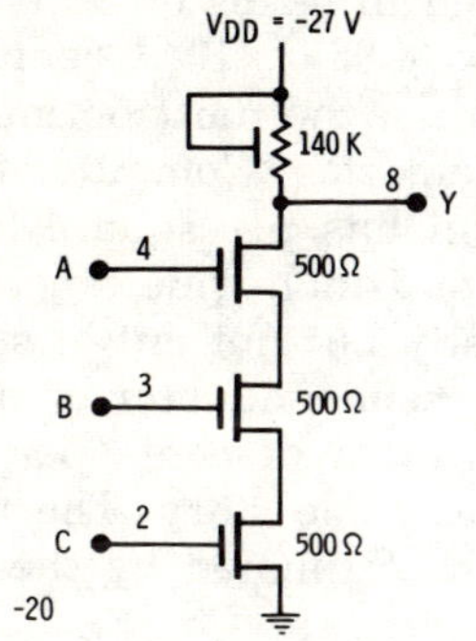

(A) NAND gate.

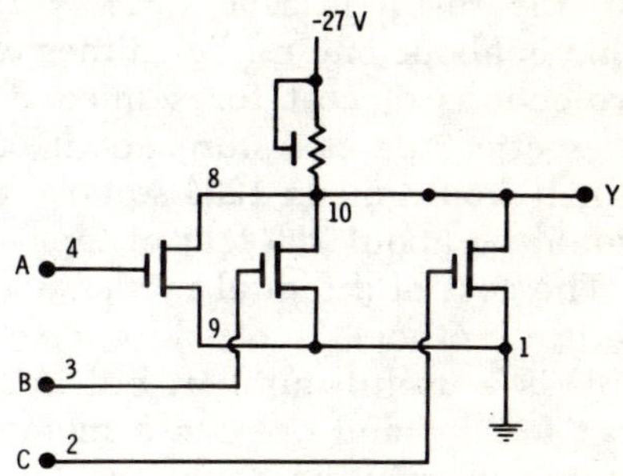

(B) NOR gate.

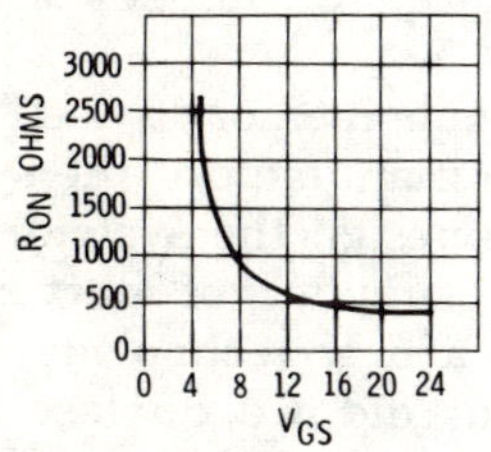

(C) Gate turn-on voltage.

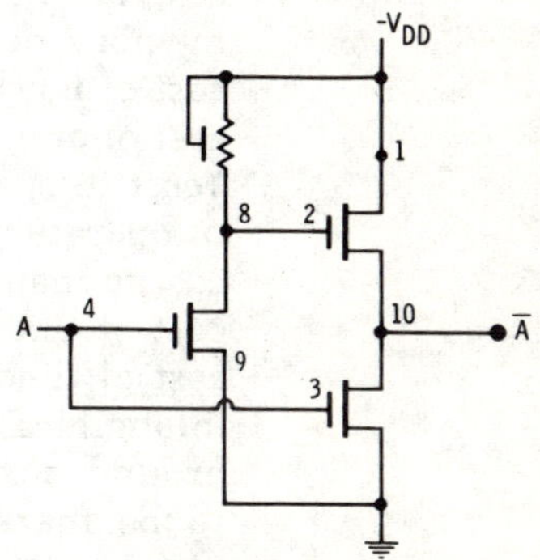

(D) Inverter-buffer.

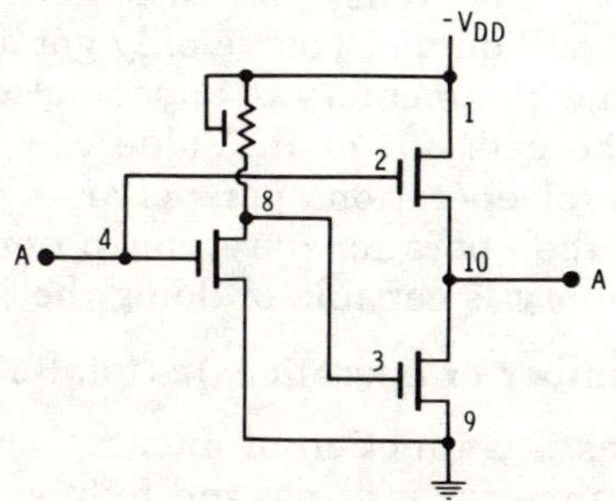

(E) Noninverting buffer.

Fig. 4-3. Three-input PMOS logic gate.

the accuracy will be one part in 1024, or about 0.124%. Therefore, we must record a 10-bit word 34,000 times each second, which gives 340,000 bits per second, or 20,400,000 bits per minute. One hour of symphonic music will be the equivalent of 1,224,000,000 or 1.224 billion bits.

An hour of music can be bought for about $5 in the form of an analog lp disc. If the cost is represented in terms of cents per bit, the result is 500¢/1.244 × 10^9 bits = 4.08 × 10^{-7} cent/bit. This is about one million times cheaper than the most optimistic projections of cost for semiconductor memories. Computer-type magnetic tape can store about one million bits per square inch, which would mean 1224 square inches. For 1-inch-wide tape, this would be about 100 feet of tape—still costly but not out of sight.

The cost of a digital memory (whether semiconductor or magnetic or other) is often represented in terms of cents per bit. This is a useful number, but it is not the whole story. The total cost to own and operate a memory can be estimated by the following descriptive equation:

Memory cost = (number of bits) × (cost per bit)
+ cost of the external support circuits that the memory needs
+ cost of power supplies
+ cost of any cooling equipment
+ (cost to operate per hour) × (number of hours of operation)
+ cost to maintain the whole memory system
+ cost of time needed to manufacture the system
+ cost of time needed to design the system
+ intangibles (Will the supplier of part X be around next year? Is a new technology ready to be marketed? What did we do last time? etc.)

The cost of a memory only tells you whether you can afford it or not. If it costs a million and you've only got a hundred thousand, forget it. The value of a memory to anyone else is determined by how much use he gets out of it. Value can be represented in terms of number of operations per dollar of total cost (assuming, of course, that the operations are worth doing in the first place and that the memory is capable of doing the job).

Memory value = (number of operations) / (total cost).

The number of operations is a function of memory access time for reading, memory cycle time for writing, and number of hours of operation.

The memory system designer, therefore, has lots of variables to play with. The value of a memory is always increased if you can make it work faster or have less downtime, or if it costs less to buy and/or operate. What may seem like a relatively small improvement in access time or cycle time may amount to a significant increase in the value of the memory.

In large memories, the cost per bit becomes the dominant term in the equation for memory cost. All the other factors get spread over so many bits that they have little influence on final cost (except reliability, which can kill you in downtime). As a result, very complex support circuits are justified if they bring down the cost per bit without at the same time significantly slowing the computer or increasing the downtime of the finished memory. Semiconductor memories are becoming more and more important because they are faster than magnetic cores, and now some types cost less per bit.

In a small memory, you want a simple, easy-to-use memory with as little support circuitry and power-supply requirements as possible. In this case, cost per bit becomes less important. In a memory designed for remote operation for a long time by means of batteries, you want low-power operation and maximum reliability.

Because of the many ways memory can be used if the cost is brought low enough, many semiconductor schemes have been developed. Dynamic PMOS memories were developed because they give low cost per bit but at the expense of complex (or at least extra) support circuits. Memories of the NMOS type give improved read and write times and operate with simpler power supplies, but originally were harder to manufacture. The CMOS memories are more expensive but allow operation with simple support circuits or none at all, while bipolar memories of various types provide very high speed.

PMOS SHIFT REGISTERS, DYNAMIC AND STATIC

Shift registers have been built in both dynamic and static versions. Dynamic registers use very little power but must be refreshed regularly or the stored data will be lost. Static registers use more power but less-complex support circuits.

The basic storage circuit of a dynamic shift register is shown in Fig. 4-4A. Capacitors C1 and C2 are the storage devices of the cell and are actually the normal input capacitance of MOS transistors Q1 and Q2. Other capacitances exist in the circuit but are not shown since they do not play any significant role in the operation.

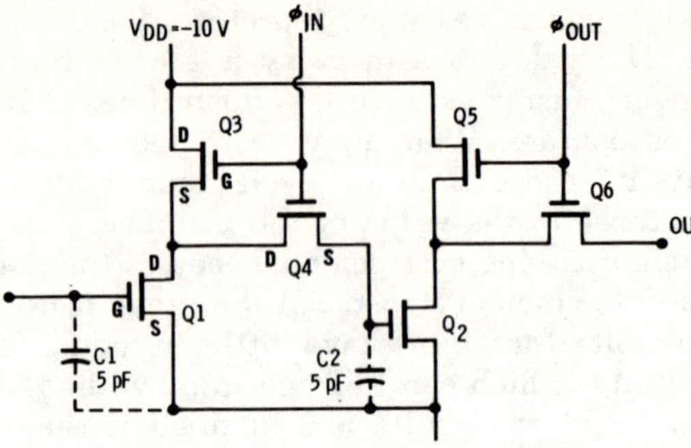

(A) Storage cell.

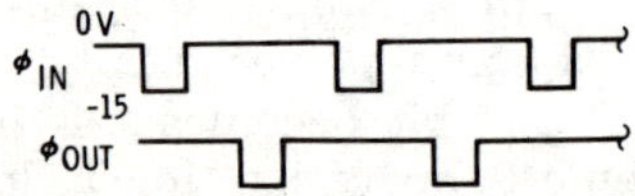

(B) Clock waveforms.

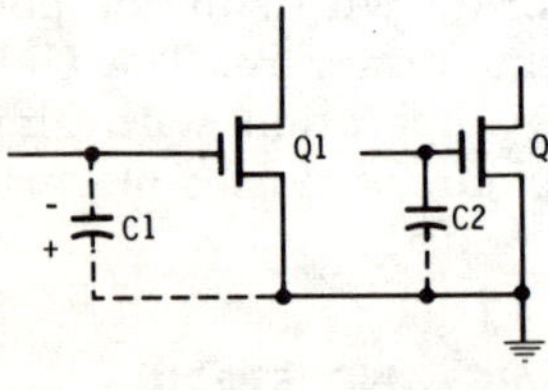

(D) Quiescent state.

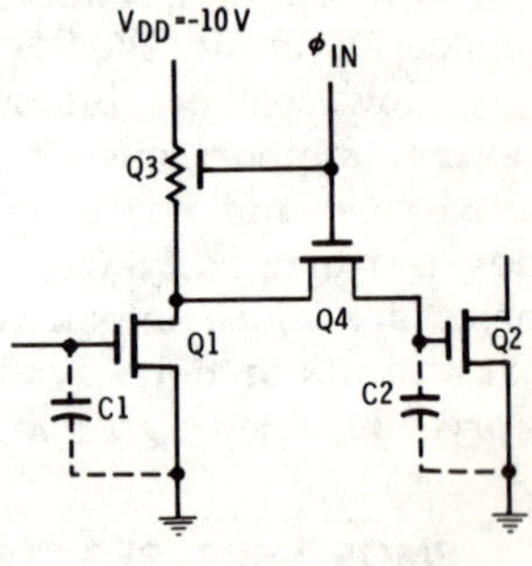

(D) Active state.

Fig. 4-4. Dynamic shift register.

In the quiescent state, both clock signals, $\emptyset_{IN}$ and $\emptyset_{OUT}$ (a clock signal is usually signified by Ø in MOS literature), are high (0 V). This causes transistors Q3, Q4, Q5, and Q6 all to be in the high-Z state. If the circuit is redrawn with these transistors omitted, it looks as shown in Fig. 4-4C. If C1 is holding a charge of the proper polarity, Q1 will be turned on; if C1 has no charge, Q1 will be off; similarly for Q2.

Let clock signal $\emptyset_{IN}$ go to −15 V, as shown by the waveform. With $\emptyset_{IN}$ at −15 V, Q3 is turned on and acts like a load resistor. Transistor Q4 also turns on and acts as a switch to connect point X to C2. Now if C1 is charged (negative voltage on the gate of Q1), Q1 will be turned on and point X will be brought to ground; as a result, C2 will be discharged. If C1 is not charged, on the other hand, Q1 is a high Z, point X approaches V_{DD}, and C2 will be charged. It is seen that Q3 and Q1 form an inverter (Fig. 4-4D) while Q4 acts as a switch; the inverter and switch are active only when $\emptyset_{IN}$ is at −15 V.

The second half of the circuit duplicates the first half, with Q5 and Q2 forming an inverter and Q6 forming a switch. In this case, the output appears when $\emptyset_{OUT}$ goes to −15 V, is an inversion of C2, and is the same as the charge that was on C1 when $\emptyset_{IN}$ went to −15 V.

On each $\emptyset_{IN}$ signal, the cell will store in C2 whatever data are at the input, either a 1 or a 0. On each $\emptyset_{OUT}$ signal, these data are transferred to the output stage and act as an input to the C1 of the next stage. Individual cells are connected into long chains to form shift registers having only a few bits or as many as 1024 bits, or more.

A simple shift register using cells of the above type is shown in Fig. 4-5. The IC is Signetics type 2506, dual 100-bit dynamic shift register with input and output inverters for buffering. The output is actually only the bottom transistor of an inverter stage and is called a bare drain output. The bare drain allows easy interfaces to other logic families and also allows the outputs of many shift registers to be connected together in the OR-tie arrangement.

The V_{CC} is specified at +5 V, the same as for TTL systems. Specifications put V_{DD} at −5 V, while the two clock signals swing from +5 to −11 V. With these voltages, the required inputs are +3.2 V or more for logic 1 and less than +1.05 V for logic 0. This gives outputs of +5 V and 0 V. If V_{CC} is set at 0 V, V_{DD} would be −10 and the clocks would swing from 0 to −16 V. For these voltages, a logic 1 input would be −1.8 to 0 V and a logic 0 input −4.4 V or lower, but not lower than −10 V.

A similar device but with additional control logic built onto the chip is type 2524, a 512-bit dynamic shift register, shown in Fig. 4-6. To read out data, the Read control is brought to +5 V. As the output clock, $\emptyset 1$, goes low, G1 is satisfied and frees G2 to follow the last bit in storage: Y is 0 if the last bit is 0 and is 1 if the last bit is 1. When Read is low, Y is low.

To write data in, Write is brought to +5 V. Gate G3 then is satisfied when $\emptyset 2$ goes low and frees G4 to follow the data. When

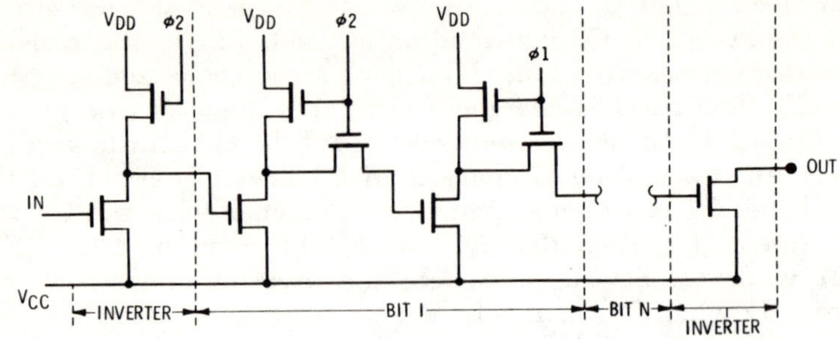

(A) Circuit.

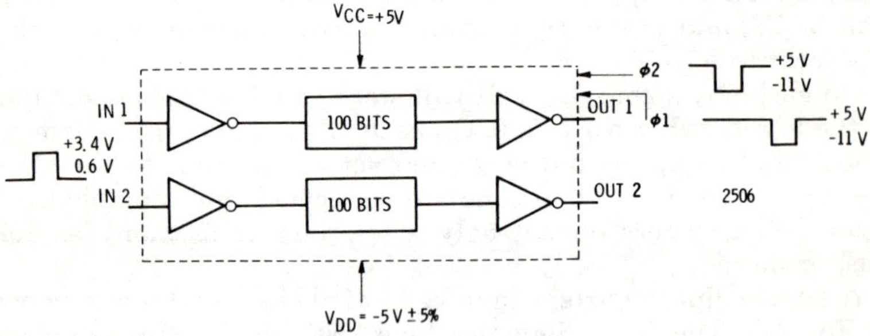

(B) Block diagram of logic.

Fig. 4-5. Dual, 100-bit, PMOS, dynamic shift register.

G3 goes high, it drives G6 low, which in turn drives G7 low, leaving G5 free to follow G4. New data will therefore be written into the register with each clock signal. Note that Read can be high simultaneously with Write, allowing both reading and writing at the same time. The bit being written in does not appear at the output immediately but only after 512 clock cycles occur. (One clock cycle equals one pulse of Ø1 plus one pulse of Ø2, usually referred to as a 2-phase clock.)

When Write is 0, G4 blocks input data while G7 and G5 follow the last bit in the register. In this mode, the data stored in the register are continuously recirculated at the clock rate. In a system using 8-bit words, the 512-bit register could hold 512/8 = 64 words. Within the register, these words are indistinguishable from one another. To pick out the beginning of any specific word, therefore, requires external circuitry, either to detect flags associated with various words or to keep track of the words by an address system, as in Fig. 4-6B. The counter keeps track of the location of bits in the register. The address bits specify the word desired. When the comparator detects a match between counter and address, it allows either the desired word to be read out or a new word to be written into the specified location. When used in this manner, the system forms a scratch-pad memory.

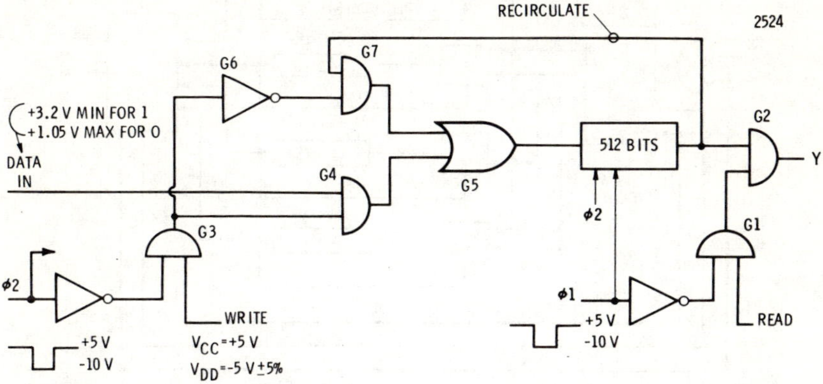

(A) Register logic.

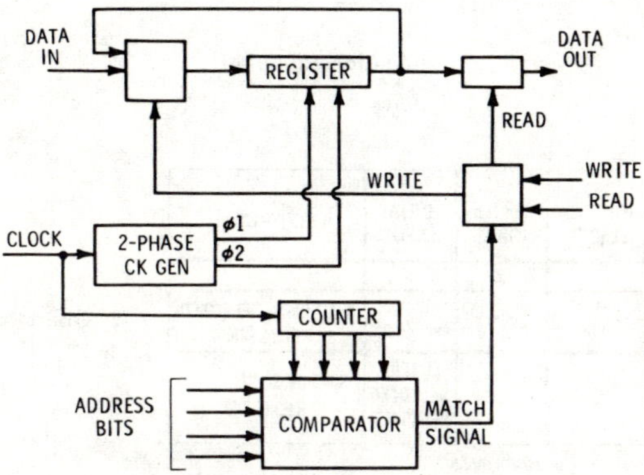

(B) Shift register as scratch-pad memory.

Fig. 4-6. Recirculating, PMOS, dynamic-shift register.

Multiplexed Shift Register

Shift registers can be combined in many ways to store data and make it available on command. Type 3325, made by Fairchild Semiconductor, and shown in Fig. 4-7, consists of four 64-bit registers with associated input and output control logic.

When the chip-select signal is high, the input decoder and the output are disabled; the output goes high and stays there while the data in the four registers recirculate continuously in accordance with the 2-phase clock. Data in the registers will carry flags

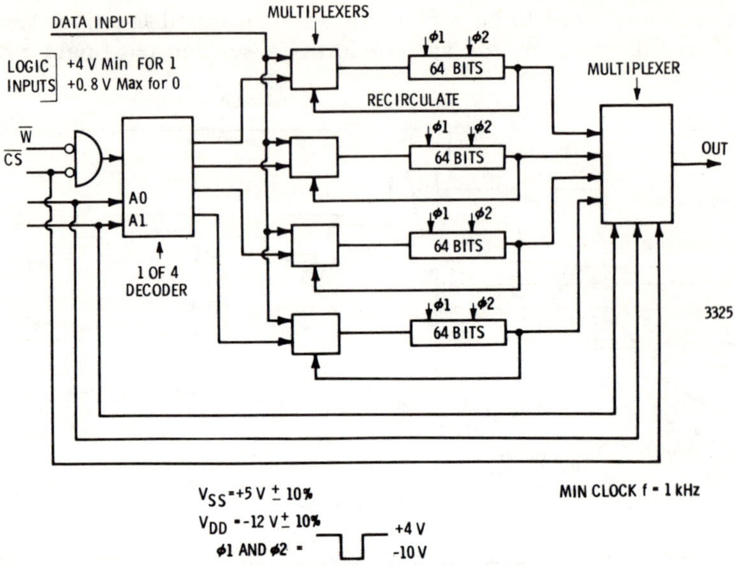

(A) Block diagram.

$\overline{CS}$ (CHIP SELECT)	WRITE INPUT	DATA INPUT	OUTPUT
H	X	X	H
L	H	X	FOLLOWS SELECTED REGISTER
L	L	ENTERS SELECTED REGISTER	FOLLOWS SELECTED REGISTER

X = IRRELEVANT

(B) Operation.

Fig. 4-7. Multiplexed, quad, 64-bit, PMOS, dynamic-shift register.

or be tracked by external circuits. When the chip-select signal goes low, the output multiplexer is enabled and one of the four registers will be selected by whatever signals exist on the address input lines. Data from the selected register are fed out through the output multiplexer while all four registers continue to operate in the recirculate mode.

If the Write input is brought low while the chip-select is low, new data will be written into the selected register. The other three unselected registers will continue to operate in the recirculate mode.

Bipolar Compatible Input

The inputs of many PMOS registers require a greater logic swing than is available from bipolar logic such as DTL or TTL. To bring the guaranteed output of such circuits to the range required by PMOS, some PMOS ICs use an extra input transistor as a pull-up resistor. When the gate of Q1 (Fig. 4-8) is connected to ground, and when other circuit voltages are as shown, the circuit will operate correctly when fed by bipolar gates. When the device is being fed by another PMOS circuit having compatible signal levels, the gate of Q1 should be tied to V_{SS}; this effectively turns Q1 into an open circuit, and thus Q1 is not seen by the input signal. The output signal, V_O, is suitable for DTL, TTL, or other PMOS devices of the same type.

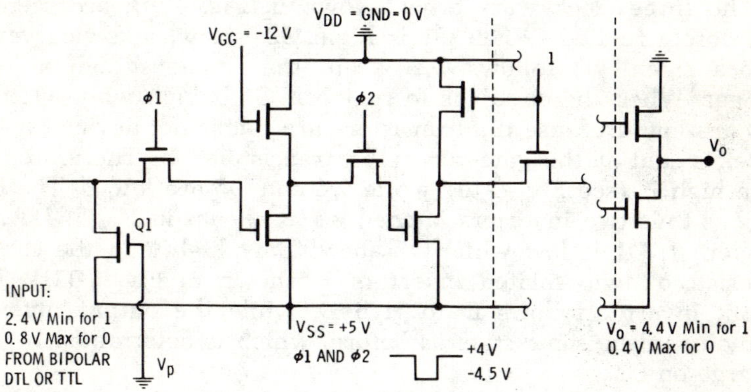

Fig. 4-8. Bipolar compatible inputs and outputs.

Static Shift Registers

If dynamic shift registers are not supplied with clock signals at a certain minimum rate depending on the specific register, the data stored will be lost. To allow operation down to zero clock rate (or dc), static shift registers were developed. These can store data indefinitely, provided normal power-supply voltages are maintained within specified ratings.

The circuit for a typical cell used widely in static shift registers is shown in Fig. 4-9A. At first glance the circuit appears to require not one clock signal but three, Ø1, Ø2, and Ø3. This is true for normal operation but not for the case where the cell is storing data in the static mode. Transistors Q2 and Q6 act as loads for Q3 and Q7, respectively, while Q1, Q2, and Q4 act as switches. Thus, the circuit consists of two inverters which can be connected in several ways.

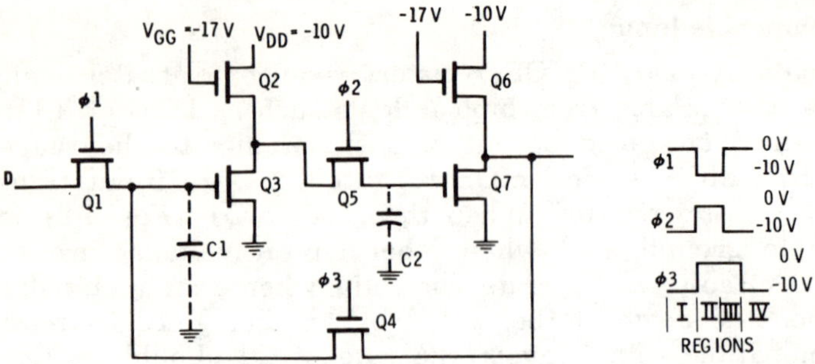

(A) Cell circuit.

(B) Clock waveforms.

Fig. 4-9. Static PMOS shift register cell.

The three clock waveforms, shown in Fig. 4-9B, are seen to be closely related. When Ø1 is high, Ø2 is low, and vice versa. Clock signal Ø3 follows Ø2 except that it stays high a little longer. When the circuit is in region I, Ø1 is high while Ø2 and Ø3 are low. Because the transistors are p-channel devices, a low clock signal at the gate turns the transistors on. Therefore, Q1 is a high Z (see Fig. 4-10) while Q4 and Q5 are low Z. The circuit is thus two inverters latched up as shown in Fig. 4-11A. In region II, Ø1 is low while Ø2 and Ø3 are high, and the circuit consists of two isolated inverters as shown in Fig. 4-11B. The input inverter follows D to give $\overline{D}$, while the output inverter remains in the same state as before, which is determined by the charge on C2.

(C) -OPERATING MODES

REGION	φ1	φ2	φ3	Q1	Q4	Q5
I	H	L	L	Hi Z	Lo Z	Lo Z
II	L	H	H	Lo Z	Hi Z	Hi Z
III	H	L	H	Hi Z	Lo Z	Lo Z
IV	H	L	L	Hi Z	Lo Z	Lo Z

Fig. 4-10. Operating modes for register cell of Fig. 4-9.

When the clock signal changes to region III, the inverters are connected as shown in Fig. 4-11C. The output goes to D. When the clock moves to condition IV, the inverters latch up again as shown in Fig. 4-11A (IV is the same as I).

A typical IC using the above cell is Signetics 2511, shown in Fig. 4-12, which is a dual, 200-bit, static shift register. Only one external clock signal is required since the three clock signals used by the cell are generated on the chip. The output circuit is 3-state; the circuit provides on-chip recirculation but does not have separate addressing circuits. When the recirculate input goes high, data are written into both registers at the clock rate.

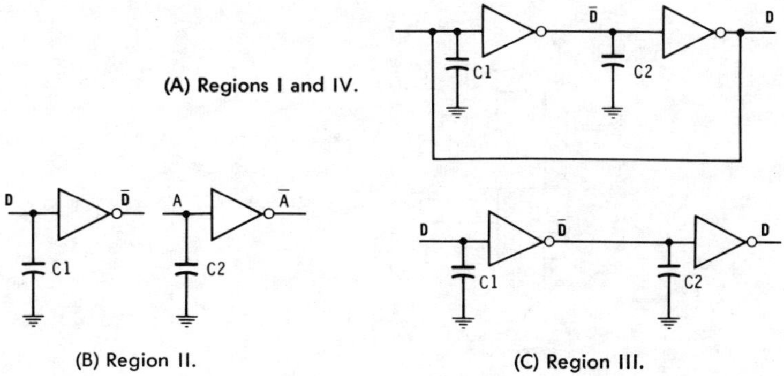

Fig. 4-11. Inverter operation of register cell shown in Fig. 4-9.

The chip includes MOS transistors to pull up inputs, including data and clock, to V_{CC} (+5 V), thus giving TTL compatibility.

The same basic circuit is also available with 50-bit and 100-bit registers.

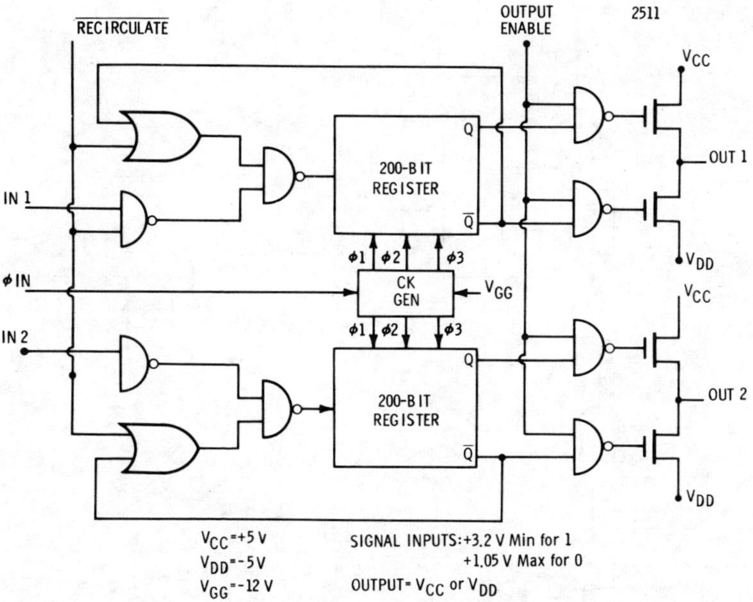

Fig. 4-12. Dual, 200-bit, PMOS, static shift register.

Variable-Length Shift Register

A variable-length shift register, shown in Fig. 4-13, can be used as a variable digital delay line, to generate pulses of various

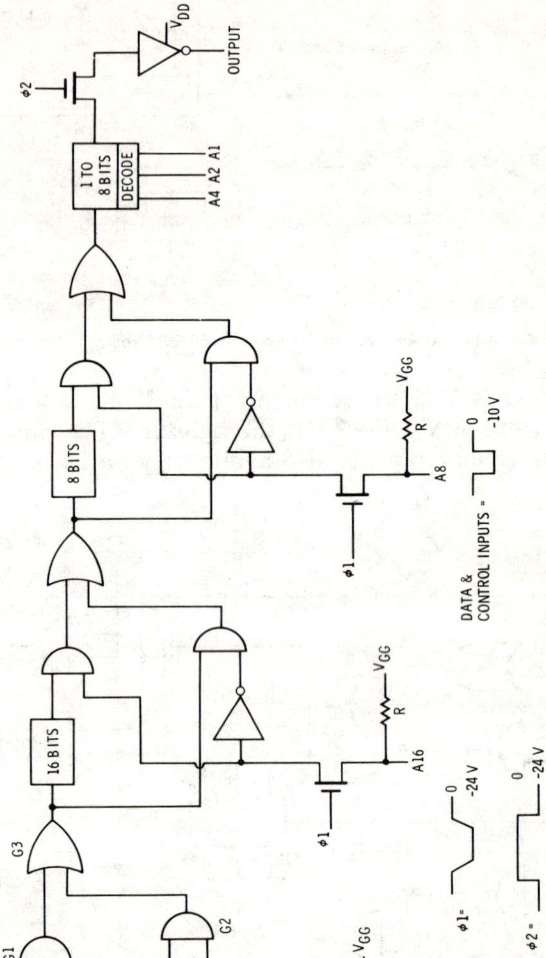

Fig. 4-13. Variable-length, PMOS, dynamic shift register.

widths or to make registers of unusual length. When Ø1 turns Q1 on, data are clocked into the register and will either enter a 32-bit section or bypass it and be fed to the input gates of a 16-bit section.

If input A32 is on, Q2 is turned on by Ø1 and G1 has one input 0 and is inhibited, while G2 has one input = 1 and is free to follow the data input. As a result, the data entering the register are shifted around the first 32-bit section and to the input of the 16-bit stage. If A32 is not on, G2 is inhibited and G1 is free to follow the output of the 32-bit section.

The same logic is repeated for the 16-bit and 8-bit stages. The final 8-bit section is controlled by A4, A2, and A1. These three inputs are decoded to provide additional storage of 1 to 8 bits. Data are released from the register when Ø2 goes low.

By selecting the correct inputs, any delay from 1 through 64 bits can be obtained.

STATIC PMOS RAM

While shift registers perform the memory function, they do not allow immediate access to any specific bit or word. A true memory IC not only stores data, but also uses address bits to provide immediate access to any bit or word in storage. If the memory is the read-write type, additional inputs allow a new word to be written into any specific location when desired.

Memories of the PMOS type have been designed with both static and dynamic circuits. Like shift registers, static types use less external support circuitry, while dynamic types use simpler memory cells and less power, store more data, and require more-complex support circuits.

The circuit for the cell of a static-type random-access memory (RAM), shown in Fig. 4-14, is a cross-coupled flip-flop. If Q1 is on, point X will be near V_{CC}, Q2 will be held off, and point Y will be near V_{DD}. If Q2 is on, Y is near V_{CC}, Q1 is off, and X is near V_{DD}.

Access to the cell is provided by Q_A and Q_B. When the cell is selected, Q_A and Q_B are turned on and output lines A and B go to the same voltage level as X and Y—one high and one low. Thus, the data can be read out of the cell. If new data are to be written into the cell, lines A and B are forced to the new state—one high and one low—Q_A and Q_B are turned on, points X and Y are forced to the levels of A and B, and Q1 and Q2 follow. When the cell is deselected, Q_A and Q_B turn off and the cell remains in its last state. As long as power is supplied to the circuit, the last data written in remain. If normal power is removed and then

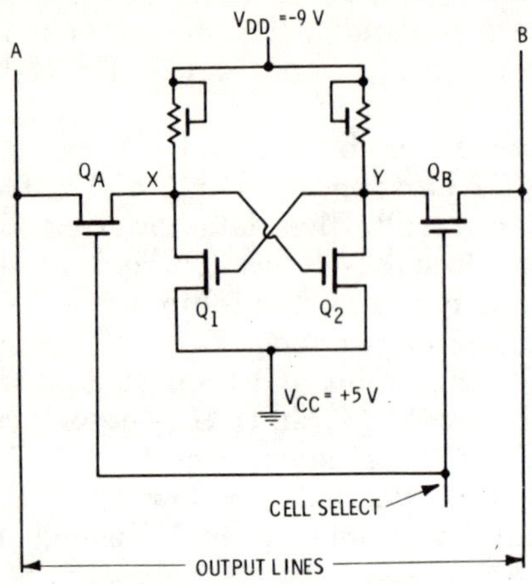

Fig. 4-14. Static PMOS memory cell.

reapplied, the levels that X and Y will take are not predictable. The data will be lost and the memory is said to be volatile.

Type 2501, one of many similar RAMs that use the memory cell of Fig. 4-14, is organized as shown in Fig. 4-15. Memory cells are arranged into 16 rows by 16 columns, giving a 256-bit memory. To select one specific bit, four address bits are decoded to select one of 16 rows, and four other address bits are decoded to select one of 16 columns. The bit at the intersection of the selected row and column is selected. If $\overline{R}/W$ is low, the selected bit is sensed and appears at outputs D and $\overline{D}$. If $\overline{R}/W$ is high, the signal at the Data In line is written into the cell.

Before considering the memory circuit in more detail, note several other features of the IC. It has a chip-select input, which, when high, disables the Data In and $\overline{R}/W$ inputs and the data outputs. When a memory array consists of a number of these ICs,

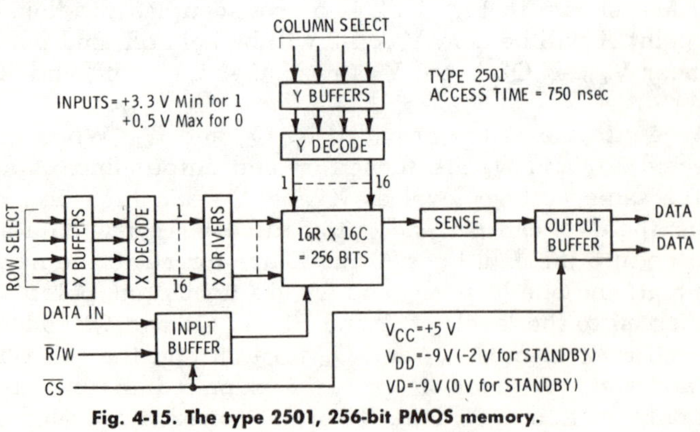

Fig. 4-15. The type 2501, 256-bit PMOS memory.

the outputs can be OR-tied together onto one common data bus and just one IC can be selected to be active at one time. The IC can also be operated in a standby mode to reduce normal power consumption. To operate in Standby, V_{DD} is reduced to −2 V and V_D is reduced to 0 V.

Fig. 4-16 shows how a four-cell version of the RAM would look, and Fig. 4-17 shows some of the control logic. When the CS input is high (+3.3 V, Fig. 4-17—note that ground is set at +5 V), Q1 and Q2 have gates and sources at nearly the same voltage

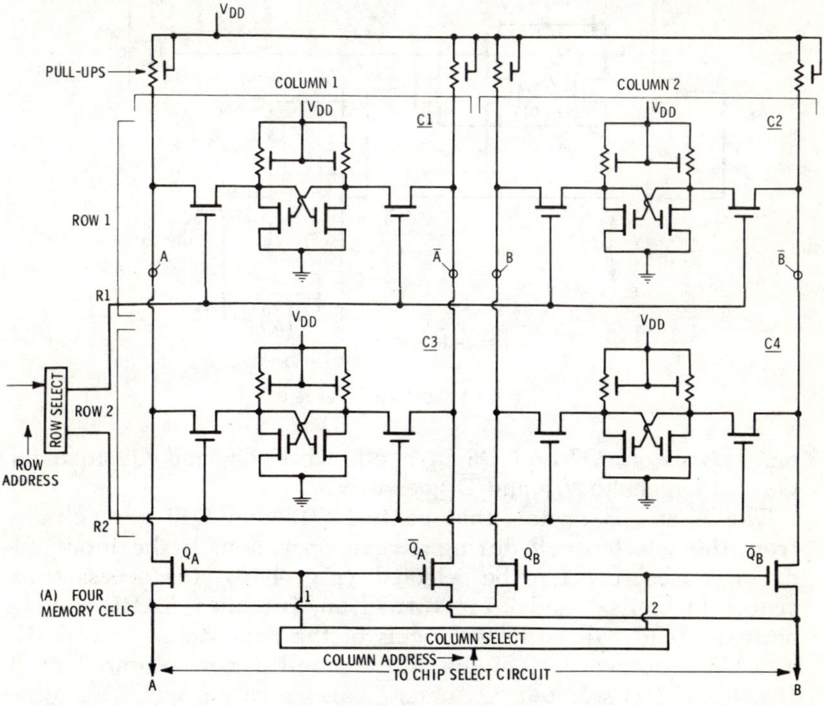

Fig. 4-16. Four-cell version of 256-bit static memory.

and are cut off; Q3, however, is on and supplies V_D to Q4 and Q5, turning them on. As a result, points 2 and 3 are brought to V_{CC}, turning off output transistors Q6, Q7, Q8, and Q9 and allowing the output to float. Note also that if V_D is at its standby level of 0 V, the outputs are also off—floating.

When CS is low (about 0 V), Q1 and Q2 are turned on. This turns Q3 off and grounds the gates of Q4 and Q5, also turning them off. Transistors Q10 and Q11 now follow Q12 and Q13, one of which will be on and the other off. If Q12 is a low Z, point 2 will go to V_{CC}, turning Q11 off and thereby causing Q10 to turn

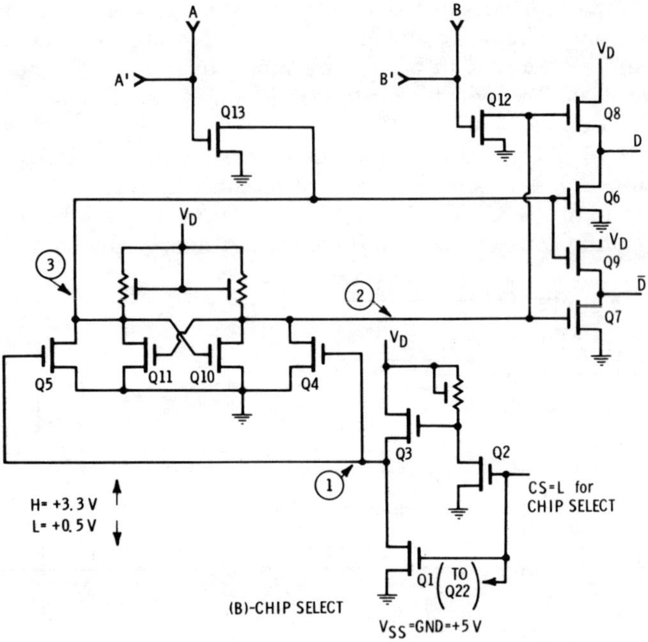

Fig. 4-17. Chip-select circuit.

on. Transistors Q7 and Q8 turn off, while Q6 and Q9 turn on. Output D goes to V_{CC} and $\overline{D}$ goes to V_D.

The A and B signals that control Q12 and Q13 are actually from the selected cell during a read operation. If the input address causes row 1 to be selected (Fig. 4-16), the access transistors for cells 1 and 2 are turned on. Bit lines $A\overline{A}B\overline{B}$ will be activated and will go to the levels of the data stored in cells C1 and C2. Another part of the address will select column 1 or 2. If column 1 is selected, Q_A and $\overline{Q}_A$ also turn on, while all other columns stay off. As a result, the data stored in cell 1 determine the state of Q12 and Q13 and thus the output.

To write data into cell 1, a circuit somewhat similar to the output circuit is used, except that now the new data input signal forces the $A\overline{A}$ to the new desired level. When $\overline{R}/W$ is low (0 V, Fig. 4-18), for reading, Q21 is a low Z. When CS = +3.3 V, Q22 is a high Z, Q23 is a high Z, and Q24 and Q26 are low Z; points M and N are near V_{CC} and Q28, Q29, Q30, and Q31 are high Z. Thus, A′ and B′ do not control Q12 and Q13.

When $\overline{R}/W$ = +3.3 V (and CS = 0 V), for writing, Q22 is a low Z, Q21 is a high Z, and Q23 is a low Z. Then Q24 and Q26 are low Z. If D_{IN} is +3.3 V, Q23 is a high Z, point N is near V_D, Q27 is

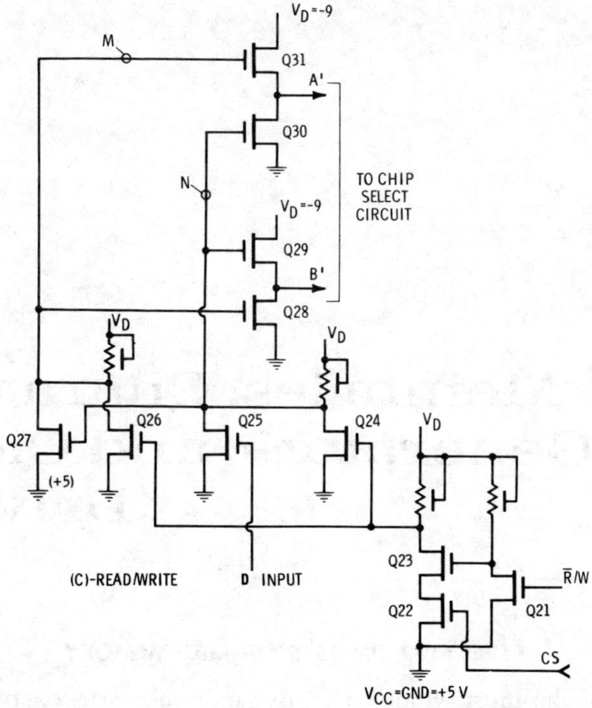

Fig. 4-18. Read/write circuit.

a low Z, and point M is near V_{CC}. For N near V_D and M near V_{SS}, Q31 and Q28 are on, while Q30 and Q29 are off. As a result, A′ goes to near V_D while B′ goes to near V_{SS}. Lines A′ and B′ thereby control lines A and B and cause the data to be written into the memory cell. If D_{IN} is low, Q25 is a low Z; then N is near V_{SS}, Q27 is a high Z, and M is near V_D; for this condition A′ goes to V_{SS} while B′ goes to V_D.

5

Memories, Character Generators, and Clock Drivers

TYPE 1103 PMOS DYNAMIC MEMORY

One of the most widely used dynamic memories is type 1103, a 1024-bit RAM developed originally by Electronic Arrays. The IC is an industry standard and is widely second-sourced, both in the original version and in many improved versions designed by a number of companies, including the originator.

The memory provides 1024 words of 1 bit each and is organized as shown in Fig. 5-1. Note that V_{SS} is specified as +16 V and V_{DD} as 0 V; thus, 0 V is negative with respect to V_{SS}, and a negative-going signal turns on a p-channel transistor. Memory cells are arranged in 32 rows by 32 columns. To select one bit, a 10-bit address is required. The first five bits of the address, A0 through A4, select one of the 32 rows, and the last five bits, A5 through A9, select one of the 32 columns. Either the bit at the intersection can be read out or new data can be written in.

A four-cell version of the memory is shown in Fig. 5-2. The basic dynamic memory cell for the standard 1103 uses three transistors. Data are stored in CM, which is the gate capacitance of Q1. To read data out of the cell, some fairly complicated and precise timing is required.

The read cycle involves a number of stages. During most of the read cycle (Fig. 5-3), the $R/\overline{W}$ input is high, which causes Q10 to be off and Q6 and Q12 to be on. In the first stage, the precharge (P) goes low, which causes Q4 to turn on to charge the Write Data line to V_{DD}; at the same time, P′ goes to V_{DD} to turn on Q5 and precharge the Read Data line.

In the next stage, CE goes low. Assuming row R1 has been selected by the address bits, point A is near V_{DD}, QR1 turns on, and V_{DD} is applied first to the Row Select line and then after a short delay, during which $\overline{CE}$ is developed and QW1 is turned off, to the Write Select line of row 1. Thus, Q2 is turned on first, followed shortly by Q3, for all the memory cells of row 1.

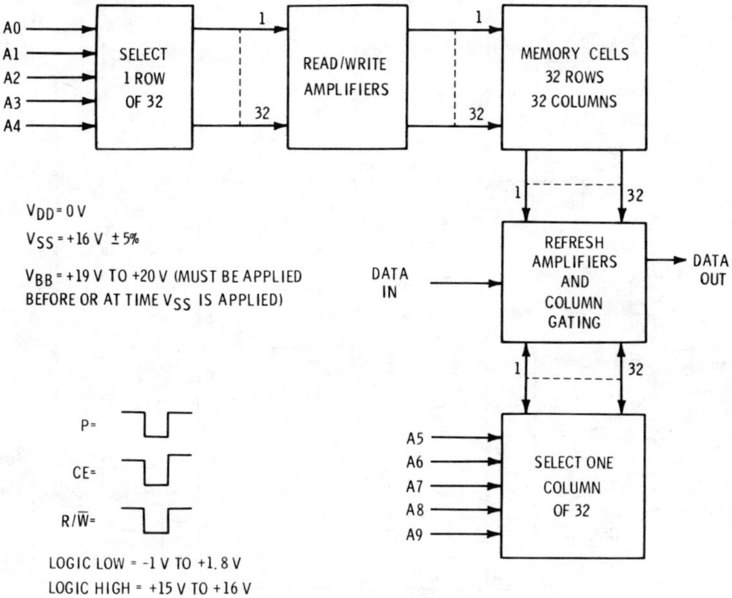

Fig. 5-1. Type 1103, 1024-bit, PMOS, dynamic memory.

In cell R1-C1, if CM is charged, Q1 turns on and the precharge on the Read Data line is shunted to ground via Q1 and Q2. Transistor Q7 remains off, and the precharge on the Write Data line is applied via Q3 to reinforce—refresh—the charge on CM. If CM was not charged, Q1 does not turn on and the Read Data line holds its precharge. This turns Q7 on, which causes the precharge on the Write Data line to be shunted to ground via Q6 and Q7. Any charge that might have accumulated on CM is also removed.

The same operation occurs in cell R1-C2, and its associated capacitor can be charged or discharged and is refreshed independently but in the same manner and at the same time as cell R1-C1. Thus, a whole row of cells is refreshed at a time.

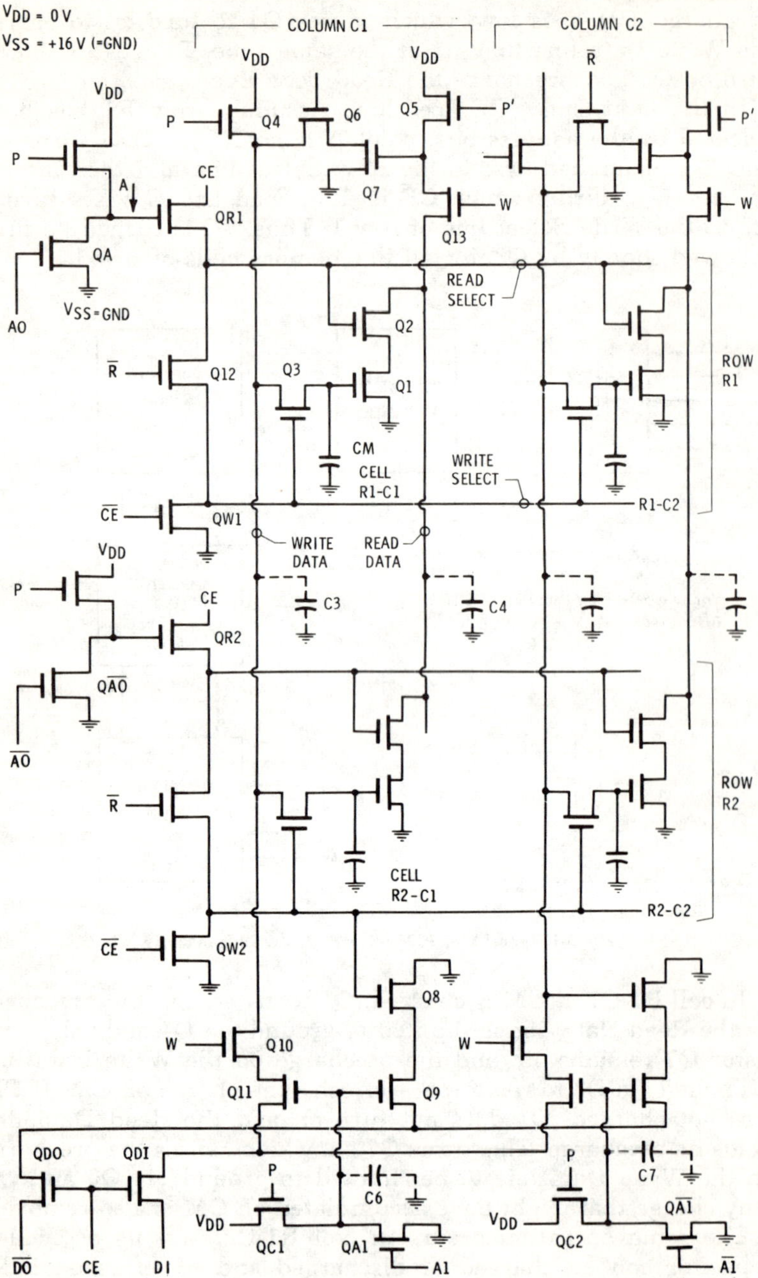

Fig. 5-2. Four-cell PMOS dynamic memory.

The signal that remains on the Write Data line is applied to Q8. If Write Data is charged, Q8 turns on. During precharge, capacitor C6 is also precharged. Assuming that column C1 has been selected, QA1 will be a high Z, which allows Q9 to turn on,

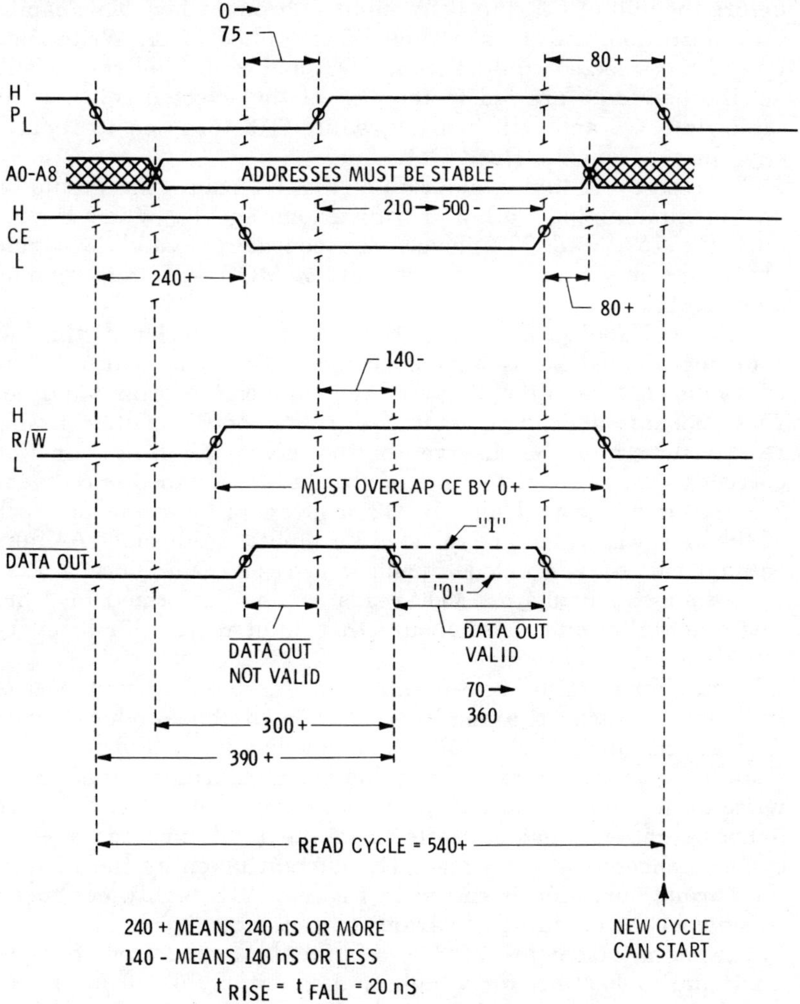

Fig. 5-3. Read cycle for type 1103 dynamic memory.

while $\overline{QA1}$ will be a low Z, which acts to discharge C7 and deselect column C2. Thus, while CE stays low, QD0 and Q9 will be on; if Q8 is on, output pin $\overline{D0}$ is connected to V_{SS} and can supply current to an external load. If Q8 is not on, output $\overline{D0}$ remains a high Z and no load current can flow. These two states are interpreted by the external circuit as a high or a low.

New data to be written into the cell are applied at the DI input pin. Then a read cycle is performed as discussed above, but just before the end of CE, the R/$\overline{\text{W}}$ input is brought low. This causes Q10 to turn on, and the signal on DI is applied to the Write Data line of the selected column. Since Q3 of the selected row is still on, the new data are fed to the CM of the selected cell. At the same time, Q6 and Q12 turn off, while Q13 turns on to remove any charge from the Read Data line.

The read-cycle timing diagram of Fig. 5-3 shows the minimum and maximum times allowed for various events. When P goes low—the 10% and 90% levels are considered valid lows and highs for many timing measurements in MOS—CE must remain high for 125 ns or more.

When CE does go low, both it and P must overlap in the low state for at least 25 ns but not longer than 75 ns. Output data are valid 120 ns after P goes high and can remain valid for 45 to 380 ns, depending on the duration of CE. Other timing relationships must be observed or the memory will not function correctly. Each row of the memory must be refreshed once every 2 ms. This is accomplished by 32 read operations, one for each of the 32 rows, which are selected by inputs A0 through A4; any column can be addressed during a refresh read. Since a read cycle takes about 0.5 μs, 4000 reads can be performed in 2 ms. The external circuit must ensure that 32 of these address every row.

Timing for a write cycle is shown in Fig. 5-4. The first part of the cycle consists of a read operation, and this produces valid data as indicated. This is followed immediately by a write pulse. Data to be written in must be stable before, during, and after the write pulse, and the output is invalid during this time. Since an extra operation must be performed, the total time for a write cycle is longer than for a read. The current taken by the IC during normal operation is shown in Fig. 5-5. It is highly nonlinear, because of the charging of various capacitances.

The output from the 1103 is a current ranging from 600 μA minimum to 4000 μA for a high, and essentially 0 (10 μA maximum) for a low. These signals must be fed to external amplifiers for conversion to suitable logic-level voltages.

A high output current corresponds to a logic 0 stored in the cell, and a zero current is a 1. The current can be changed to a TTL level with the circuit shown in Fig. 5-6. For a logic 0 in the cell, the voltage across the 3K resistor will range from $3 \times 10^3 \times 600 \times 10^{-6} = 1.8$ V to $3 \times 10^3 \times 4 \times 10^{-3} = 12$ V. However, the

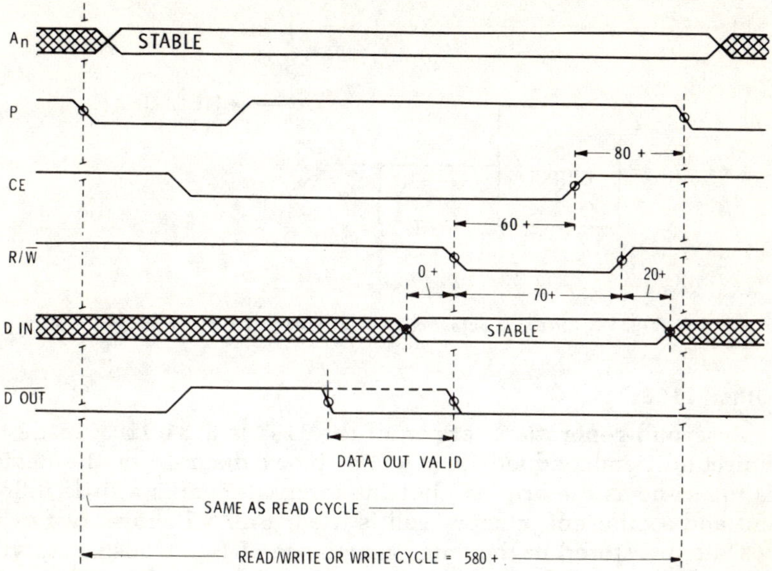

Fig. 5-4. Write cycle for type 1103 dynamic memory.

output impedance of the memory will limit the voltage to a much lower value. Outputs can be OR-tied together to produce large memory arrays.

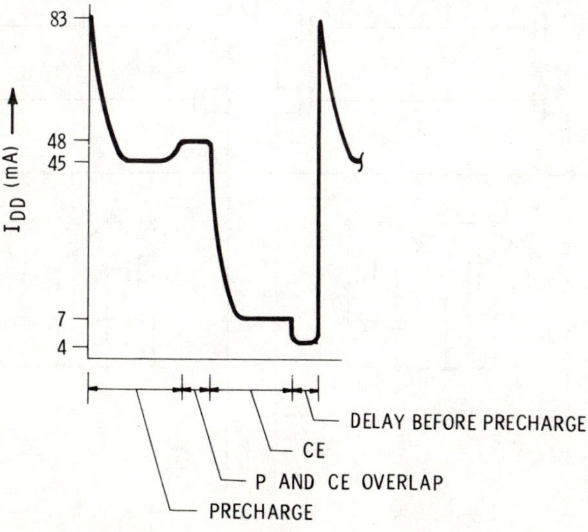

Fig. 5-5. Curve of I_{DD} current in type 1103.

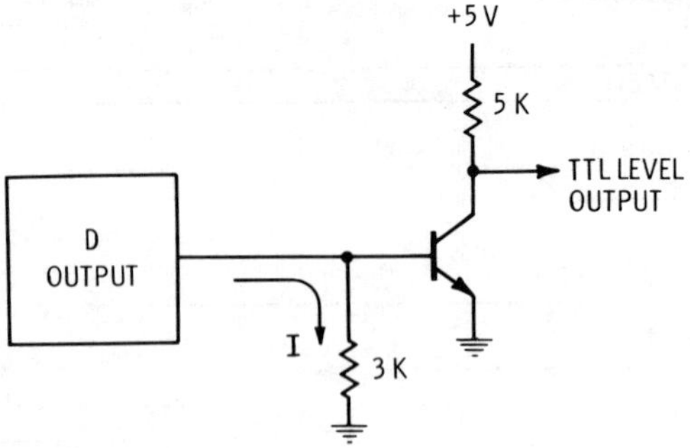

Fig. 5-6. Changing output currents of 1103 to TTL level voltages.

Other 1103s

A second-generation version of the 1103 is 3534/1103, made by Fairchild Semiconductor Corp. The block diagram of the device is the same as the original, but the internal logic is a little different and a different memory cell is used; Fig. 5-7 shows two cells.

Data are stored in the gate capacitance of two transistors, with two other transistors providing access. In the quiescent state, either C1 or C2 will be charged to near V_{DD}, and thus either Q1 or Q2 will be in the high-Z state. Since the charge on C1 or C2

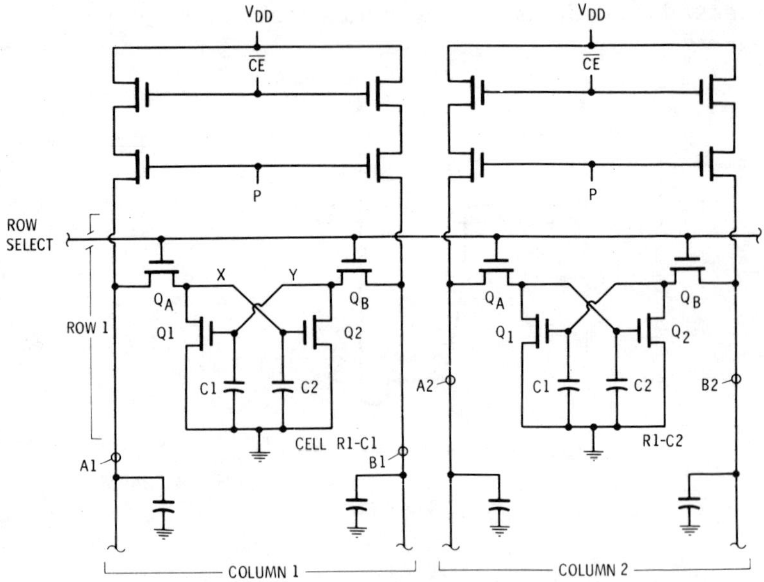

Fig. 5-7. Two cells of 3534/1103 dynamic memory.

will leak off fairly soon, the data must be refreshed, and the circuit is dynamic.

To read data out of the cell, the A and B lines are first charged to V_{DD} by applying precharge signal P, and these charges are stored in parasitic capacitances C3 and C4. The row-select line is then brought to V_{DD}, and Q_A and Q_B are turned on. If Q1 is on because of a charge on C1, point X will be near ground and the Data line and C3 will be discharged through Q1 to ground, and any charge that might have accumulated on C2 is removed at the same time. Since Q2 will be a high Z during this time, point Y will be near V_{DD} and the charge on C1 is refreshed.

Sensing and column-select circuits determine whether the A line is high or low. While data are read out of the cell at the row-and-column intersect, the whole row of cells has been refreshed. Thus, 32 reads suffice to refresh all 1024 bits.

To write information into cell R1-C1, line A1 is brought to one state (high or low) and then Q_A and Q_B are turned on; C2 is either charged or discharged by this action, and C1 and B1 are driven to the opposite state. When Q_A and Q_B are turned off, C1 and C2 hold their respective charges until the next cycle.

The timing diagram for a read cycle is shown in Fig. 5-8. The most significant difference occurs in the relationship of P and CE. The first-generation device requires that these two signals

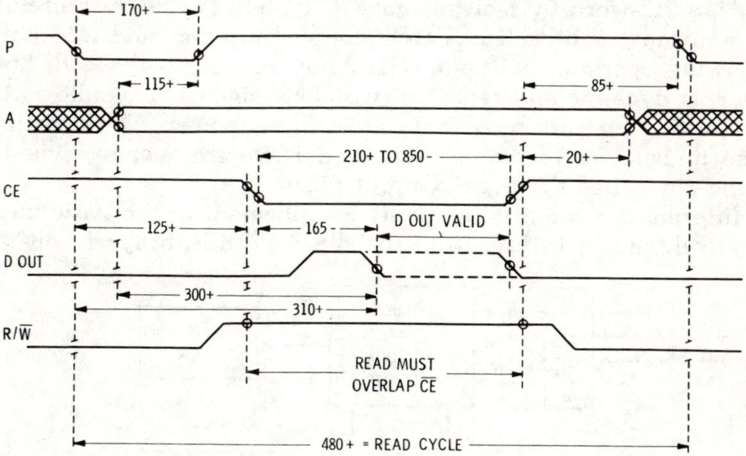

170 + MEANS 170 nS OR MORE
165 - MEANS 165 nS OR LESS

$t_{RISE} = t_{FALL} = 20$ nS

Fig. 5-8. Read cycle of 3534/1103 dynamic memory.

overlap by at least 25 ns but not more than 75 ns. Also, once CE starts toward low, P must return high within 25 to 140 ns. The second-generation device does not require any overlap but will tolerate it, although it does require that CE go low no sooner than 125 ns after P goes low.

Other versions of the basic 1103 have been developed, primarily to increase operating speed. These are summarized in Table 5-1. Other objectives include simpler support circuitry and reduced power consumption.

Table 5-1. Type 1103 PMOS Memories

Industry Part Number	Access Time (μs)	Cycle Time (μs)
1103	300	580
1103S146	220	350
1103-1	180	300
1103-2	150	250
3534/1103	300	480

STATIC/DYNAMIC PMOS RAM

A cross between a static RAM and a dynamic RAM is the 3532, a 512-word by 1-bit memory (Fig. 5-9), made by Fairchild. The memory cell itself is a cross-coupled flip-flop, and it is static, allowing operation without refreshing. Access to the cell, however, is dynamic and requires two clock signals. This allows the supporting circuits to operate at reduced power. The circuit is also unusual in that data enter and leave the memory via the same pin, called the Input/Output pin.

Internal operation is relatively complicated. In the basic memory cell, shown in Fig. 5-10, the flip-flop will always be in one

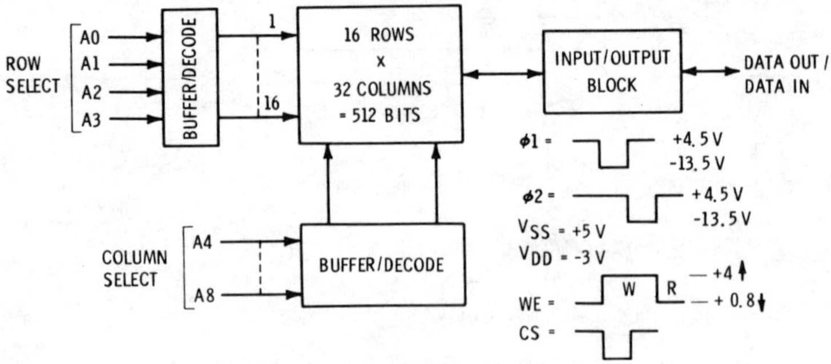

Fig. 5-9. Static PMOS memory with dynamic readout.

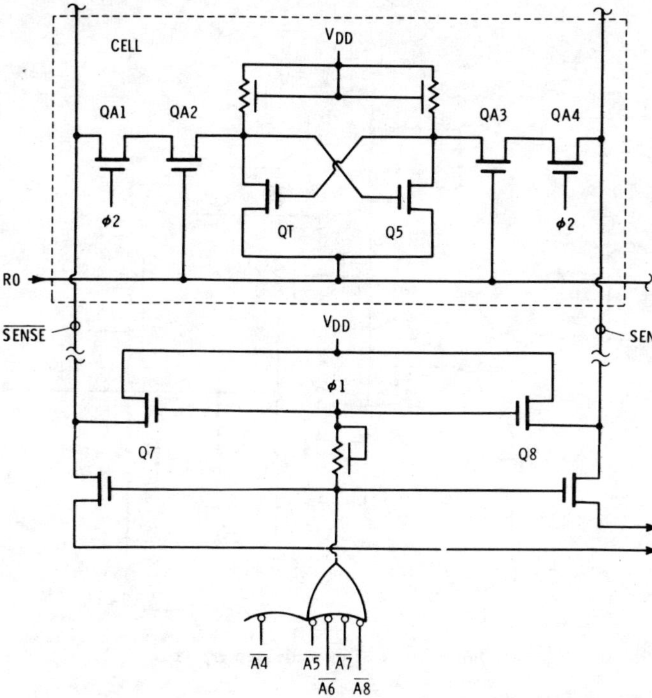

Fig. 5-10. Memory cell of PMOS memory with dynamic readout.

state or the other and QA1, QA2, QA3, and QA4 provide access. A memory cycle, either read or write, begins when Ø1 goes low. After a short delay, as shown by Fig. 5-11, the signal levels at the eight address inputs determine which cell is to be selected.

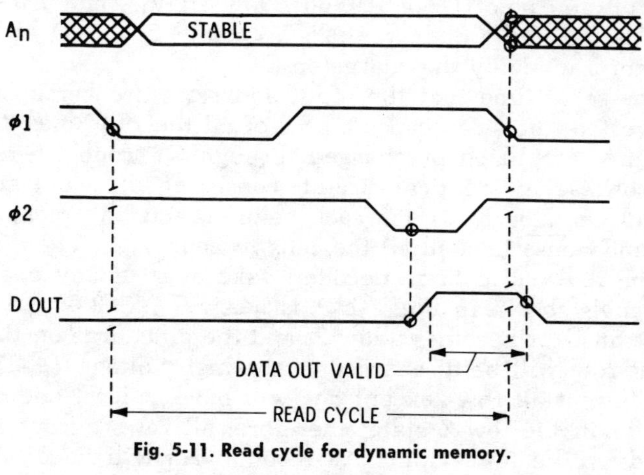

Fig. 5-11. Read cycle for dynamic memory.

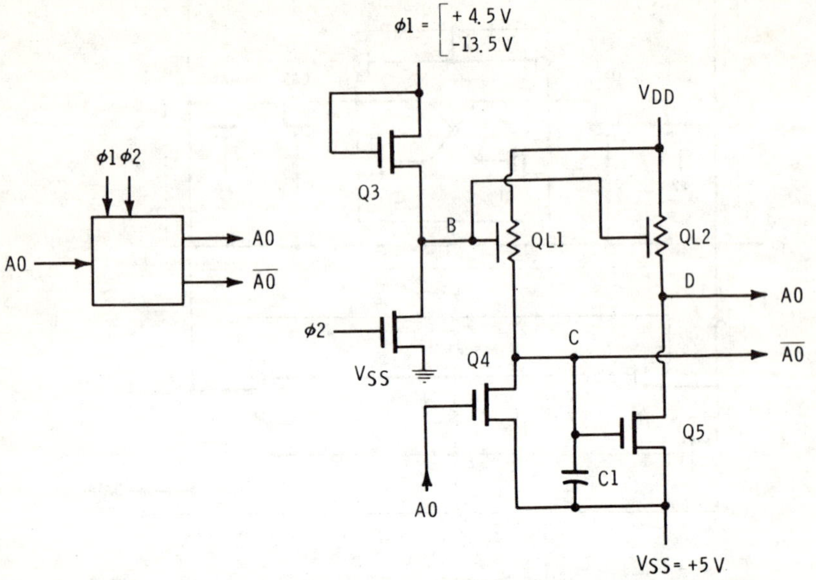

Fig. 5-12. Input circuit of dynamic memory.

Each input is fed to an inverter/amplifier (Fig. 5-12), which generates signals An and $\overline{An}$. While Ø1 is low (−13.5 V), Q3 is a low Z and point B is about −13.5 V, which activates the two load transistors, Q_{L1} and Q_{L2}. If input A0 is low (that is, more negative than V_{SS}, or about 0 V, since V_{SS} = +5 V), Q4 turns on and point C is brought near V_{SS}. Transistor Q5 is therefore turned off and point D, or A0, goes toward V_{DD} and $\overline{A0}$ goes to V_{SS}. When Ø1 goes high, to +4.5 V, points C and D are maintained for a while by the charge on C1.

At the same time that the input addresses are being buffered and inverted, the stray capacitances of all the row lines, CRO in Fig. 5-13A, are being precharged through Q6 (one Q6 for each of 16 rows). An identical circuit, consisting of two transistors (Q7 and Q8, Fig. 5-10) for each column, acts to precharge the Sense and $\overline{\text{Sense}}$ lines in all the cells to near V_{DD}. When Ø1 goes high, the row and column decoders take over. If any one of the gate signals shown in Fig. 5-13A is near V_{DD} (−3 V), the associated transistor becomes a low Z, and the precharge on the CRO for that row will be discharged to the high voltage (+4.5 V) of Ø1. Note that all rows except one will have at least one decoder transistor in the low-Z state; therefore, all rows except the one selected by the four inputs A0 through A3 will be discharged.

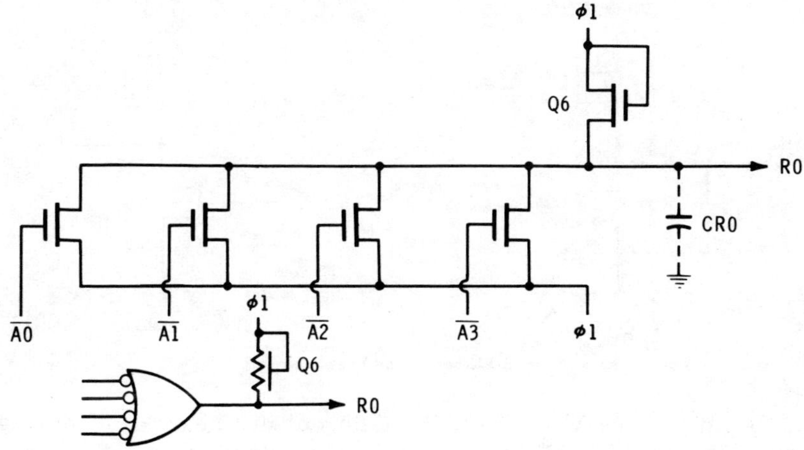

(A) Row decoder.

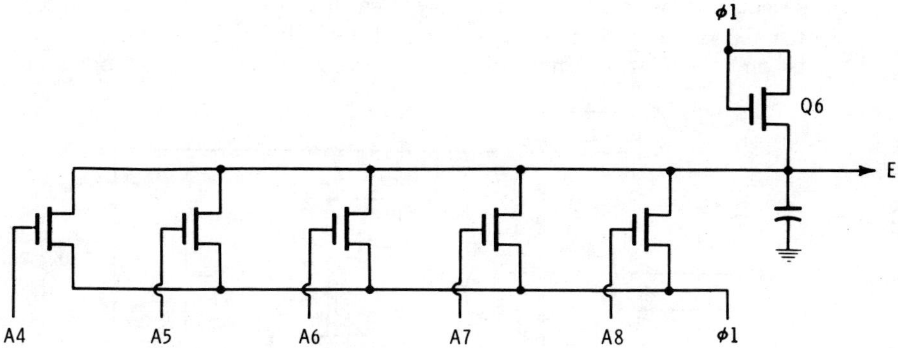

(B) Column decoder.

Fig. 5-13. Decoding systems.

Similarly for the columns shown in Fig. 5-13B—all except the one selected by inputs A4 through A8 will be discharged.

The chip-select signal now determines circuit operation during the rest of the cycle. If the chip has not been selected, the CS input will be +4.5 V, Q13 (Fig. 5-14A) will be a high Z, and the internal $\overline{CS}$ line will have been charged to −3 V during Ø1. For a read cycle, the write-enable input, WE (Fig. 5-14B), will be near 0 V. In the Input/Output circuit (Fig. 5-15), the WE signal acts to disable AND gates G1 and G2 while G3 is free to follow the $\overline{Data}$ line. The Ø1 signal is also supplied to Q11, which causes C11 to become charged and turns Q_O on, bringing the

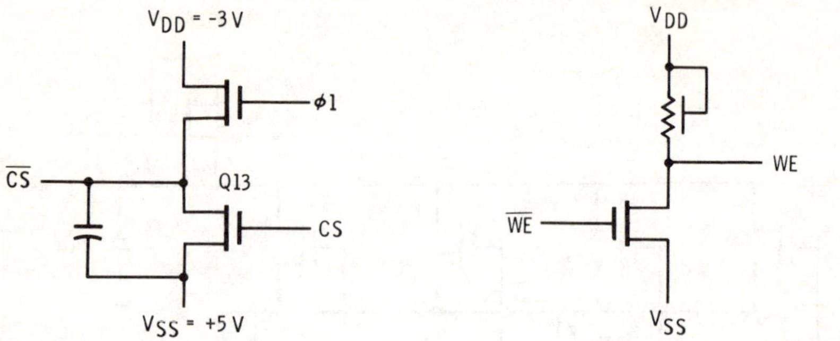

(A) Chip select. (B) Write enable.

Fig. 5-14. Select and enable circuits.

I/O pin to near V_{DD}. Thus, the data out are near −3 V during this Ø1 period and are not following the data in the cell.

After the output circuit and the row and column lines have stabilized, Ø2 can go low. The cell at the intersection of the selected row and column now has all four access transistors turned on. If QS is on, the Sense line will be discharged. If QT is on, the $\overline{\text{Sense}}$ line will be discharged. The Data and $\overline{\text{Data}}$

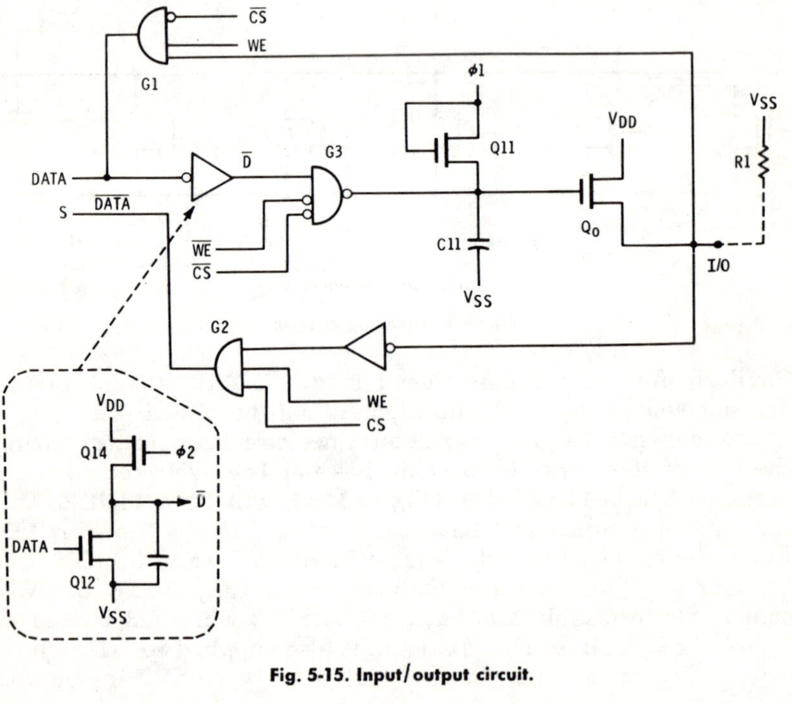

Fig. 5-15. Input/output circuit.

lines from the column will thus stay high or go low, depending on the data stored in the cell. If the Data line stays charged (Fig. 5-15), Q12 stays on, $\overline{D}$ stays low, G3 does not respond, and the I/O line stays low. If the Data line from the cell being accessed is discharged, Q12 is turned off, and $\overline{D}$ is brought near V_{DD} via Q14. Gate G3 is thereby turned on and discharges C11, causing Q_O to turn off. As a result, the output pin is pulled to V_{SS} via external resistor R1.

When Ø2 goes high, the circuit goes to the standby mode. But the output data remain valid for several microseconds, until the charge on C11 leaks off. A short time after Ø2 goes high, the circuit is ready for a new cycle.

The write cycle (Fig. 5-16) is similar to the read cycle except that the WE input is high during the cycle. This frees gates G1 and G2 (Fig. 5-15), disables G3, and causes Q_O to turn off. The I/O pin must now be driven either high or low, depending on the data that are to be written into the selected cell. As a result, the Data line is driven to one state—high or low—and the $\overline{\text{Data}}$ line is driven to the opposite state. The new data thereby force the selected cell to the desired state.

The device works with external data-handling circuits of the type shown in Fig. 5-17. Outputs can be connected together in the wired-OR circuit, and common output resistor R1 allows point X to follow the I/O pin of a number of devices. Gates GO and GI then provide input and output signals for the memory.

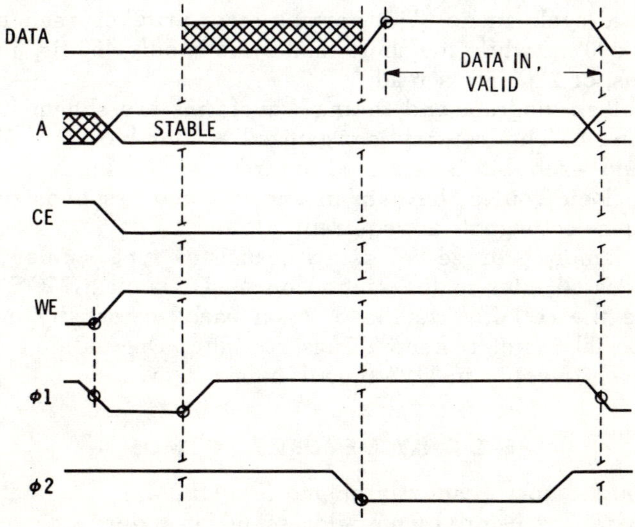

Fig. 5-16. Write cycle.

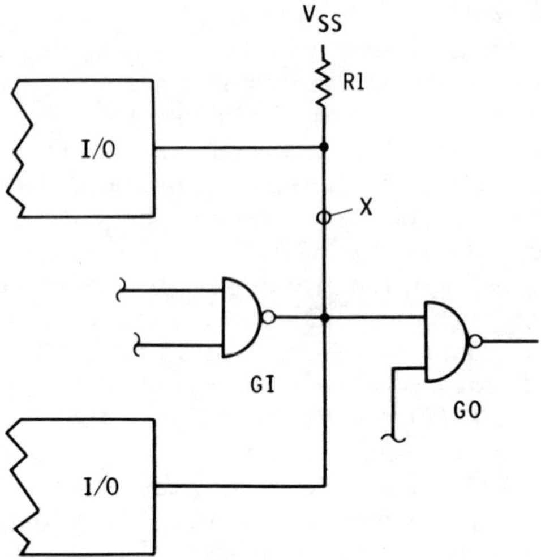

Fig. 5-17. Input/output select circuit.

External logic determines which chip is selected, whether the cycle is read or write, and when the output is valid.

Three-Phase Dynamic RAM

Another variation in dynamic RAMs is type 2548, which is a 2048-word by 1-bit memory that uses three clocks. The use of the extra clock allows the device to be read out in a special short-read-cycle mode. The normal read/write or read/refresh cycle is 590 ns, while the short read cycle is only 460 ns, a saving of 130 ns, or 22% for a read.

The block diagram and timing waveforms are shown in Figs. 5-18A and B. The memory is organized as four blocks of 512 bits each, and each bit is accessed by row and column decoders. On-chip logic routes the data in and out and responds to W/R and chip-select signals as required.

The memory storage cell is a 3-transistor type similar to the cell discussed later in the section on 4k Dynamic NMOS RAM. Because the cell inverts the data on each access, the memory uses special logic to keep things straight. The data-out signal is a current level as in 1103-type dynamic RAMs.

READ-ONLY MEMORIES IN PMOS

Although most read-only memories (ROMs) use static addressing, a few use dynamic addressing for reduced power consumption. In the static ROM, the addresses are applied and the

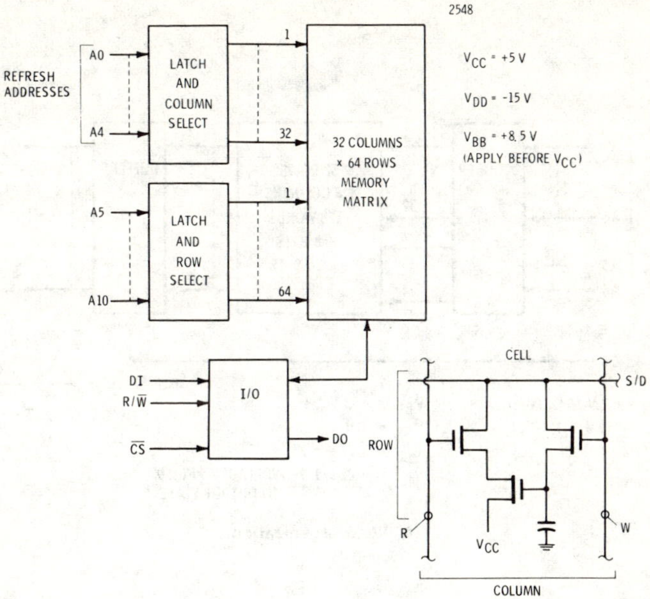

(A) Block diagram.

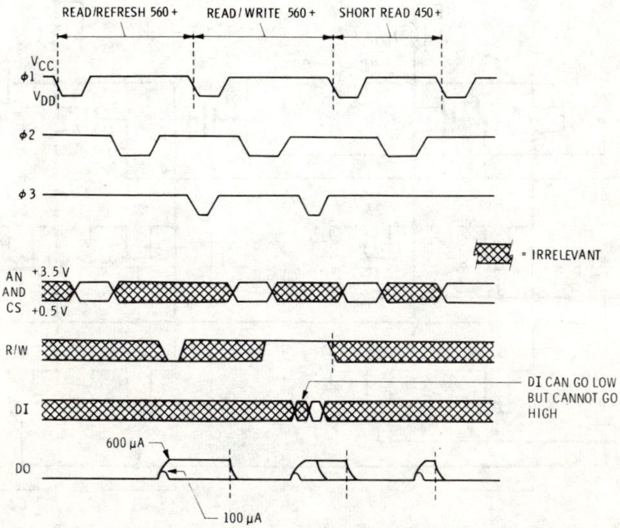

(B) Timing waveforms.

Fig. 5-18. Type 2548, a 2048-bit, 3-clock, PMOS, dynamic RAM.

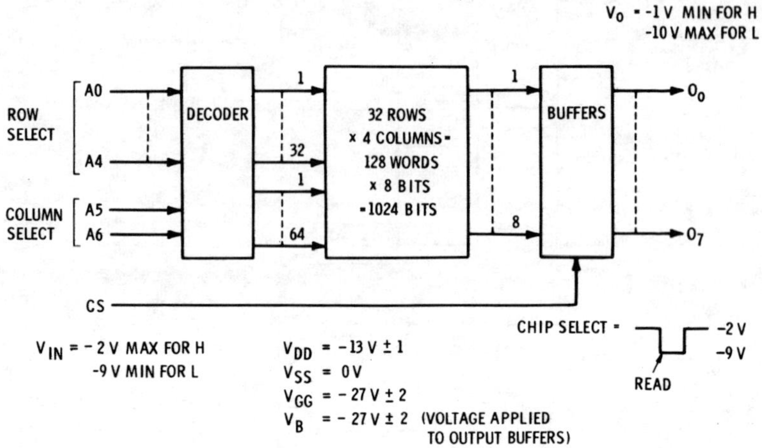

(A) ROM organization.

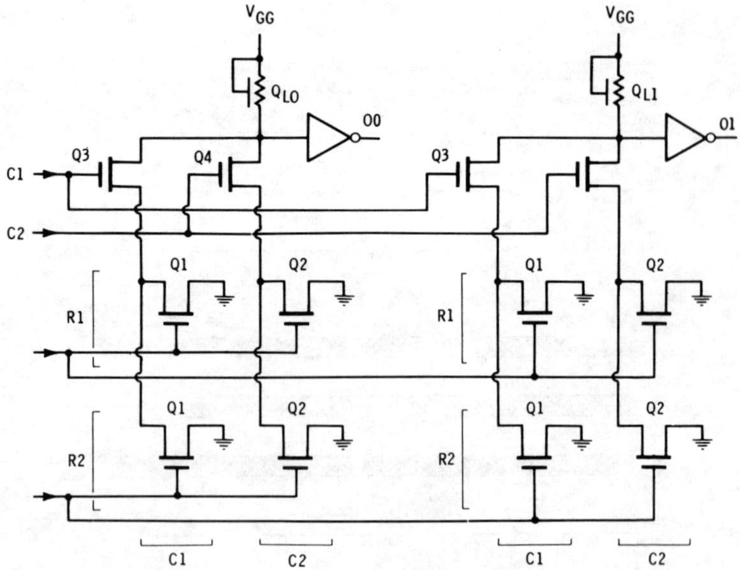

(B) Four-word by 2-bit version of the memory.

Fig. 5-19. Type 3501,

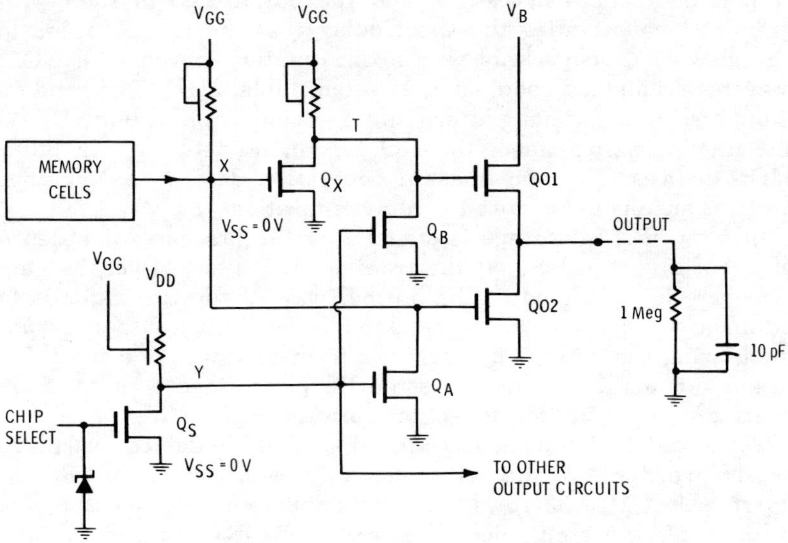

(C) Chip-select and output functions.

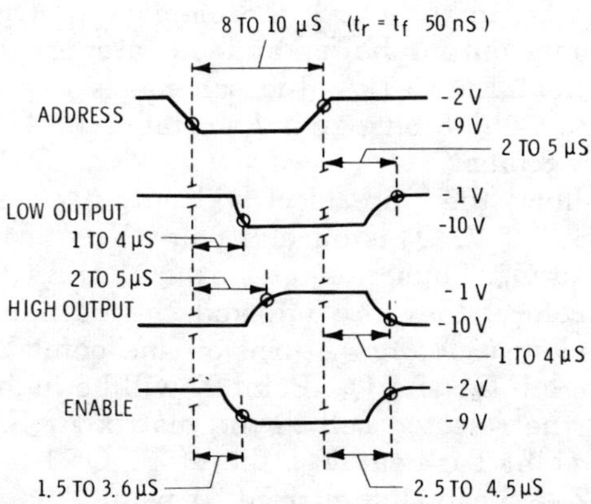

(D) Response times.

a 1024-bit, PMOS ROM.

chip is enabled, if necessary, and the outputs go to their programmed values after the short delay that the circuit needs to charge various capacitances. The information stored in a ROM may be a standard code, such as a trig table, a table of random numbers, or a special pattern for a specific application. Up to the final masking stage, the ROM is manufactured like other MOS devices. The final masking operation determines whether an H or an L is to be stored in a given location.

In some ROMs, storage is effected by the presence or absence of a working transistor. If the transistor is to be active, the gate structure is completed in the normal way. If the transistor is to be inactive, the gate structure is made ineffective, either by not metallizing the gate or by making the insulating oxide too thick to allow transistor action. In some ROMs, storage is effected by making or omitting a connection between logic gates.

Type 3501 is a typical ROM. It is a 1024-bit device organized as 128 words of 8 bits, as shown in Fig. 5-19A. Five address inputs select 1 of 32 rows, and two address inputs select 1 of 4 columns. If the chip select is low, the eight outputs go to the state stored in the memory.

The memory cell consists of one transistor. The cells are arranged as indicated by Fig. 5-19B, which shows how a 4-word by 2-bit version of the memory would look. If row 1 is selected, transistors Q1 and Q2 are turned on. If column C1 is selected, Q3 is turned on. Therefore, point X is pulled to ground through Q1 and Q3. Output O0 goes to $\overline{X}$. If a ROM is designed to store an L at R1-C1, Q1 will be inoperative and will remain in the off state regardless of its gate voltage. In this case, when Q3 is turned on, point X remains high and output O0 goes low.

For the next bit in the word, the memory matrix is repeated and an additional output buffer and pin are provided. For the memory shown at Fig. 5-19A, the scheme is expanded into 32 rows by 4 columns by 8 bits, giving a total of 1024 bits organized as 128 words of 8 bits.

Fig. 5-19C shows the chip select and output circuit of the device. When CS is 0 V, Q_S is off and point Y is near V_{DD}, which turns on Q_A and Q_B. Thus, the gates of QO1 and QO2 are pulled near V_{SS}; therefore, they are off, and the output is a high Z. When CS goes low (−9 V), Q_S turns on and point Y goes toward ground, turning off Q_A and Q_B. Point X will be high or low since it is following the selected cell in the matrix. If X is near V_{GG}, QX is on, and point T is near V_{SS}; therefore, QO1 is a high Z, but QO2 is a low Z and can sink current. If point X is near V_{SS}, QX and QO2 are high Z, while point T is near V_{GG}; therefore, QO1 is a low Z. Current can then flow from V_B through QO1 to the output circuit. For load current of 10 μA, the output voltage swings from −1 V to −10 V.

The memory is relatively slow, requiring from 3.6 to 4.25 μs for access, as indicated by Fig. 5-19D. Note also that the access time required to reach a high is different than the time required to reach a low.

PROMS AND APROMS IN PMOS

If you need a memory in which the data must be changed often, you need a read/write device or RAM. If the data do not change and must not be lost when the power is turned off, you need a ROM. If you need a ROM but do not know what data it has to hold, you need a field programmable ROM, or PROM. If you need a memory in which the data sometimes change and must not be lost when the power goes off, you need an alterable PROM, or APROM.

Field programmable ROMs are essentially ROMs that are designed and built with all transistors or connections at every bit location. The original outputs, then, are all H (or all L). To change an H to an L, the bit is selected by a typical address scheme and then an overload pulse is applied to the output pin. The overload destroys the transistor or connection, and from then on the output from that bit location will be an L. Once a bit has been changed from an H to an L, it cannot be changed back. Once programmed, PROMs function just like ROMs.

Alterable PROMs use special transistors that have two different threshold voltages. One threshold voltage is low and allows normal transistor action at normal working voltages. The other threshold is high and does not provide transistor action at normal working voltages. So far, at least two methods of altering threshold voltage have been developed. In one case, if the memory transistors are exposed to strong ultraviolet light for 20 to 30 minutes, they go to the low-threshold state. They can then be reset selectively to a high threshold by applying higher than normal bias voltages. Once programmed in this manner, they stay in this state until subjected to another ultraviolet light treatment. These types of PROMs are packaged with a top that is transparent to ultraviolet light.

The other type, APROM, uses transistors whose threshold can be changed both ways electrically. Fig. 5-20A shows the block diagram of type NC7001 (by Nitron), a 64-word by 1-bit device (NC7002 is similar but holds 1024 bits), and Fig. 5-20B shows a 4-cell version. To program the memory, all the memory transistors are first set to the low-threshold state by applying V_{DD}

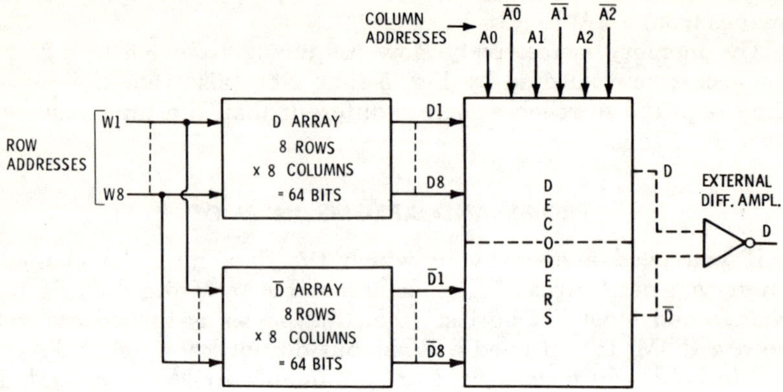

(A) Organization.

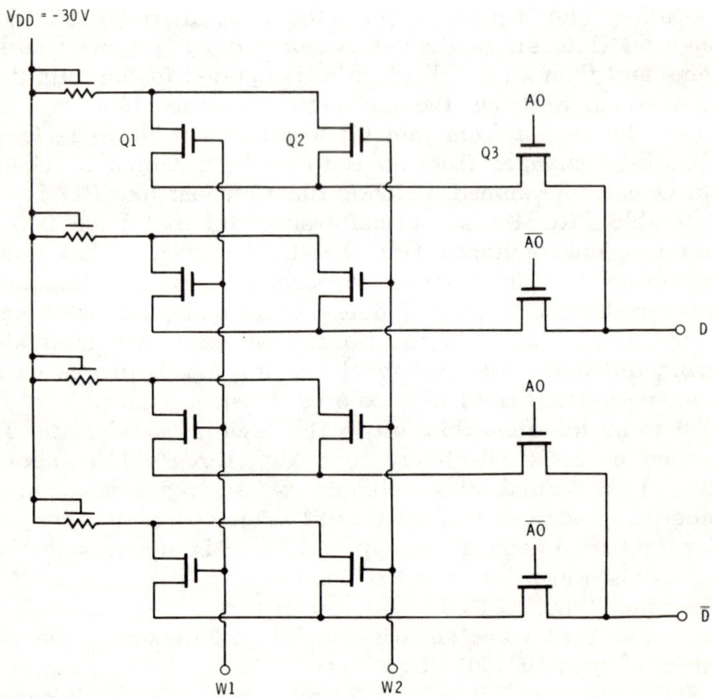

(B) Four-cell version.

Fig. 5-20. Electrically

(−30 V) to the V_{SS} terminal—the substrate—for about one second. To write data into the memory, either the D or $\overline{D}$ pin is grounded. If an H is to be written, the D pin is grounded and $\overline{D}$ left open. The row address then determines which row of memory transistors have their sources grounded. If Q3 has been selected, Q1 and Q2 will have their sources at 0 V and their drains at V_{DD}. One of the column-select lines, such as W1, is then brought to −30 V, turning on Q1. When the memory transistor is subjected to these conditions for from 1 to 10 ms, its threshold is permanently shifted in a negative direction, which means that the transistor will not turn on when interrogated later at normal working voltage levels. The transistor in the array for the $\overline{D}$ does not have its source grounded since $\overline{D}$ has been left open. As a result, this transistor remains in the low-threshold state. When it is interrogated later at normal working voltage, it will turn on. All cells in the memory are then programmed in the same manner.

The read operation is similar to the write, except that the voltage levels are different, as shown in Fig. 5-20C, and the output is a current from either the D or the $\overline{D}$ terminal. These two signals, when applied to a differential-input amplifier, give the H or L stored in the memory.

Because of the way the memory works, the inputs do not use protective networks and thus extra care is required in handling.

CHARACTER GENERATORS (PMOS)

Letters, numbers, and other characters and symbols can be generated on the face of a crt by moving the electron beam around, as in an oscilloscope, or by pulsing the beam on and off in a scanning mode, as in a tv set. Because of the ready

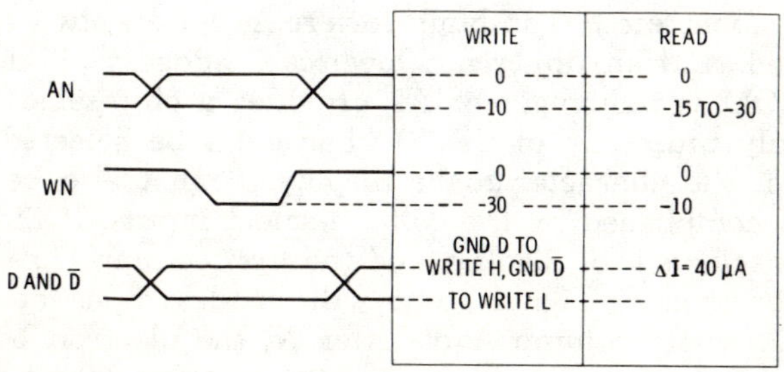

(C) Waveforms.

alterable PMOS PROM.

availability of low-cost tv sets and their adaptability as video terminals, the scan mode has become popular and a relatively large number of character-generator ICs have been developed to perform this function.

Several character fonts have become accepted as standard for these applications. Fig. 5-21A shows how the letter *F* is formed in a 5 × 7 dot matrix. As the electron beam is moved from left to right along line 1, five dots are placed on the tv screen, one in each column. On the next scan across the screen, a dot is placed in column 1 only, etc. In all, seven scans are available to write the letter *F* or any other character. The persistence of the screen holds the information between scans.

Fig. 5-21B shows how an *F* is formed in a 7 × 9 matrix. In Fig. 5-21C, 16 scans are used, but the character is written on only 9 adjacent lines. This allows a character to be raised above the normal position to act as a superscript (such as in X^2) or to be lowered to a subscript. This also allows a character to be combined with an overline or underline.

To write a sentence, such as the one in Fig. 5-21D, numerous letters, spaces, and punctuation marks must be placed on the screen in the proper relationship. The information to be written on the screen is first generated, perhaps by a keyboard or a computer, and then stored in a memory. The display system then works with the memory and control logic to generate and display the information. Information in the memory can be changing as necessary and will appear on the screen in the next frame.

Read-only memories that operate somewhat differently than conventional ROMs are used to generate the characters. A typical ROM for generating ASCII (American Standard Code for Information Interchange) 5 × 7 characters is type TMS 4103, which is organized as shown in Fig. 5-22. Character address is provided by A1 through A6, which are decoded to select one of 64 characters (A through Z, 0 through 9, plus punctuation marks, signs, symbols, etc.). To obtain lowercase letters, two ROMs are often used, and an uppercase/lowercase address bit determines which ROM is interrogated; A7 provides a chip-select function.

Not only must one of the 64 characters be selected, but the column of the character to be displayed must also be selected. This is accomplished by the column-select inputs. If the letter *N* has been selected for display, and the first column is selected, all 7 outputs produce signals that will be drawn as dots on the screen. To write column A of letter *N*, the electron beam must be moved from top to bottom (or bottom to top) and must produce a dot at each of seven locations. After column A has been written, the beam is moved to column B and the procedure is repeated, with a dot being written on row 2. To write the whole letter *N*, all five columns must be turned on in proper sequence.

In the horizontal-scan mode of writing, assume that output 1 is used to control the electron beam. The beam is then moved horizontally at a constant rate and the columns are turned on in sequence. The beam produces a dot in columns A and E. If a complete line is being written, as indicated in Fig. 5-21D, the beam will be blanked out while it is moving from the last column of *N*, through the space between the letters, to the first column of

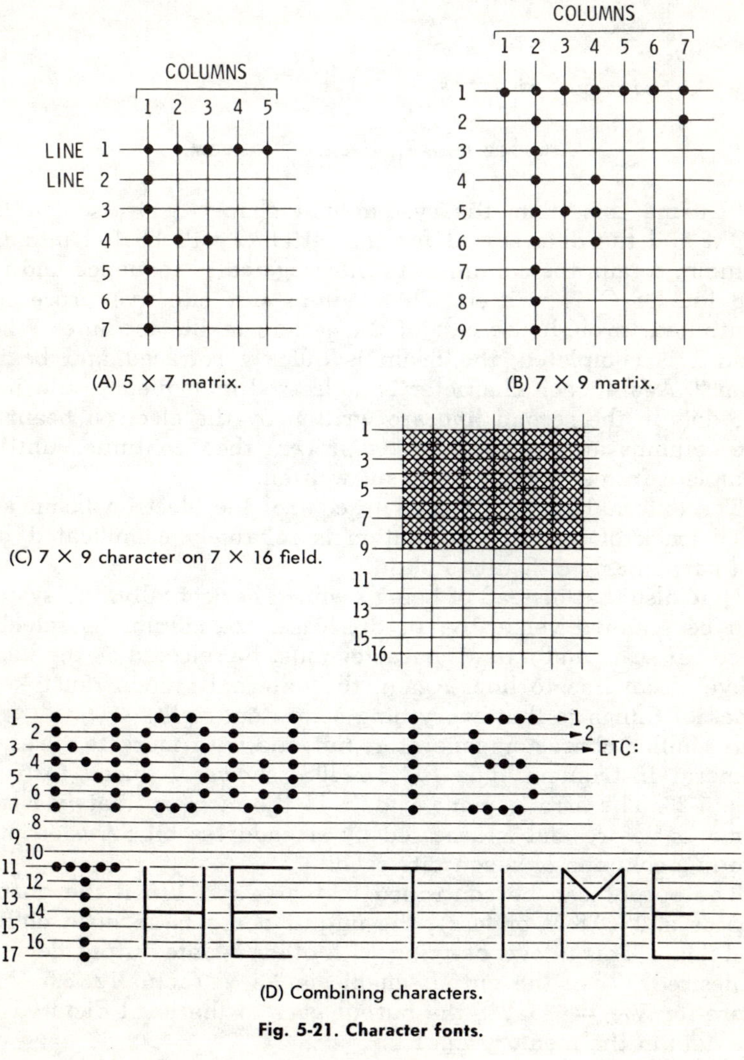

(A) 5 × 7 matrix.

(B) 7 × 9 matrix.

(C) 7 × 9 character on 7 × 16 field.

(D) Combining characters.

Fig. 5-21. Character fonts.

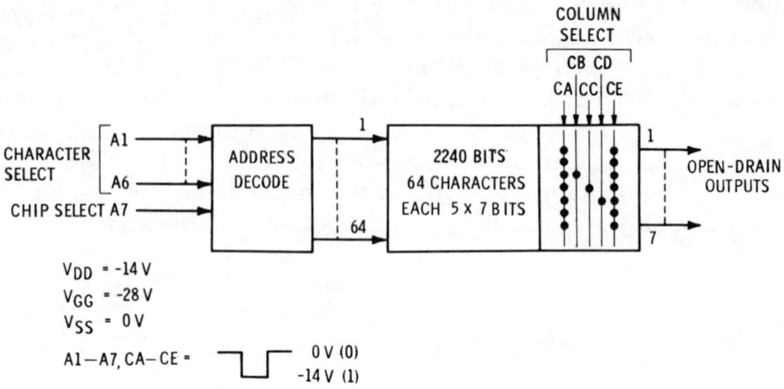

Fig. 5-22. Character generator PMOS ROM.

O. During this time, the system will change addresses to the ROM and the data stored for the letter *O* will be fed into the column. Again, the columns are interrogated in sequence and the top line of *O* is written. This tedious but effective process is continued through the rest of the letters in the top line. When scan 1 is completed, the beam is quickly retraced and begins scan 2. Again, each character is addressed in sequence, and now the dots in the second line are written by the electron beam as the columns are addressed. The process then continues until a complete frame of dots has been written.

The external system needed to control the electron beam and keep track of all the information is relatively complicated but not particularly difficult to build.

Note also that instead of being scanned horizontally, the system can be scanned vertically. In this case, the column is selected once per scan, and a new character must be selected as the beam moves from line to line. Again, the external system must keep track of things so that everything comes out right.

A similar character generator, but one with more features, is General Instrument type RO-5-2240S, shown in block form in Fig. 5-23. The arrangement of data in the memory matrix is the same as before and is accessed by six address bits; the outputs are five columns of seven bits each.

The output can be controlled in two ways. When the output enable is 0 (0.8 V or less), the output becomes an open circuit and floats. This allows OR-tying with other 3-state output devices, if desired. When the output enable is +5 V (actually, 3.5 V or more for V_{CC} = +5 V), the output goes to the level dictated by the data in the memory matrix.

The blanking input provides another level of output control. When it is +5 V, the output is determined by the memory. When it is 0 V, all outputs go to +5 V. If a stream of data is being processed by the system but a display or printout of the data is not desired, the blanking input can be used to inhibit this operation.

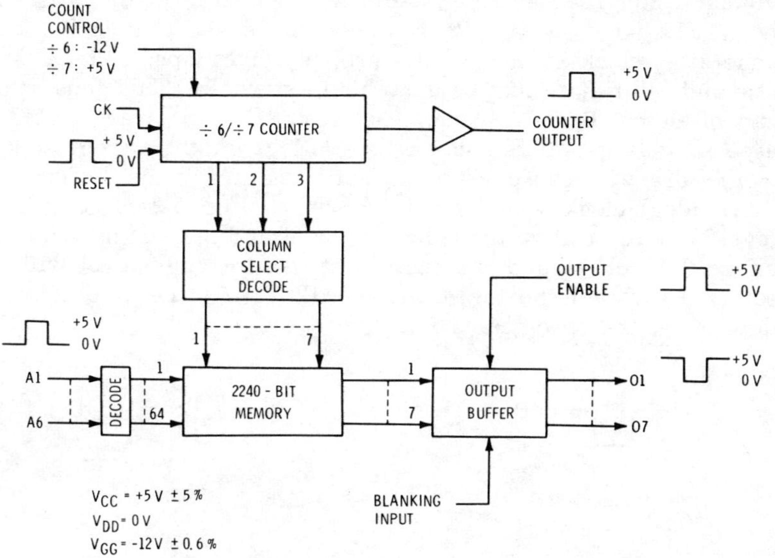

Fig. 5-23. Variable-space PMOS character generator.

With respect to column selection, only one clock input is required instead of five. A clock signal is fed to the internal counter which generates the column-select signals in sequence. In this case, the columns cannot be selected in arbitrary order but must occur one after the other. An additional feature is provided, in that the counter actually controls either six or seven columns of output information. The first five columns are data from the memory and provide the signals that are to be displayed. Columns 6 and 7 are blank, and this provides spacing between successive characters. If the count-control signal is at −12 V, the counter allows five columns of data plus one column of blanks to appear at the output; if the count-control is at +5 V, the counter allows five columns of data plus two columns of blanks.

As a further aid to integration in a larger display system, the counter generates an output signal when it reaches a full count (either six or seven), and a separate input allows it to be reset to zero.

CLOCK DRIVERS FOR MOS CIRCUITS

Some MOS devices have clock drivers built on the chip, and for these circuits all that is needed is the proper V_{GG} and V_{DD} voltages, plus the regular system clock, which in many cases can be a TTL-level signal. When the IC does not contain a clock generator, clock signals of the proper waveshape, voltage, and rise and fall times must be generated externally. This generation can often be done with simple circuits. But in larger systems, especially large MOS memories, relatively high current spikes can be drawn because of the capacitive nature of the devices.

An ideal clock waveform is shown in Fig. 5-24A. The high level is more positive than the low level and may or may not be above 0 V, or ground. In many systems, the high level will be set at +5 V and the low level at −12 V, but the actual levels depend on the specific system.

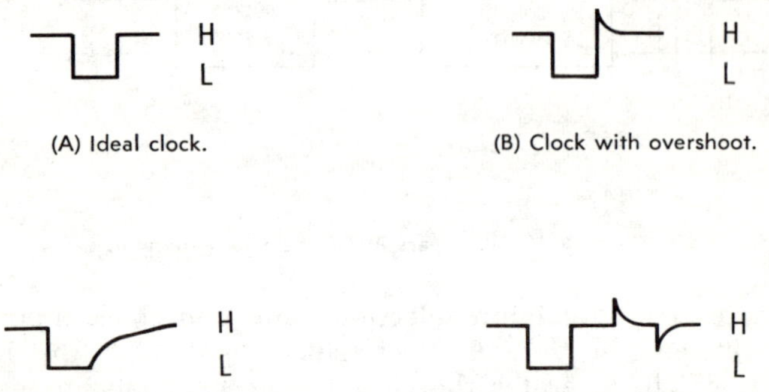

(A) Ideal clock. (B) Clock with overshoot. (C) Slow recovery. (D) Coupled noise.

Fig. 5-24. Clock waveforms.

The waveform in Fig. 5-24B has an overshoot. If the overshoot is large enough to drive the substrate diode into conduction, the IC may operate erratically and in extreme cases can be damaged or destroyed. The slow-recovery waveform in Fig. 5-24C delays the turn-off of the circuit being driven by the clock and can result in improper operation. In Fig. 5-24D, another signal, often the second phase of a 2-phase clock, is being coupled onto the normal clock signal. If the amount of noise is large enough, it can cause trouble, especially if it occurs only under certain conditions of system operation.

A number of special circuits have been developed to generate acceptable clock signals for MOS circuits. Type SH0013 by Fairchild is a dual-driver device that can deliver ±600 mA at up to 30 V. The device can be driven by a standard TTL gate, such as a 9002 or a 54/7400, to generate 2-phase clock signals. It can also be driven with improved performance by higher-power TTL buffers or TTL line drivers.

Fig. 5-25 shows one circuit of the IC being driven by a standard TTL NAND gate. The signal from the TTL gate is fed into the clock generator via a resistor and capacitor in series. This network controls the width of the pulse that will be delivered to the output circuit. Because the TTL signal is coupled capacitively, the generator circuit can operate at the voltage level needed for the clock signal.

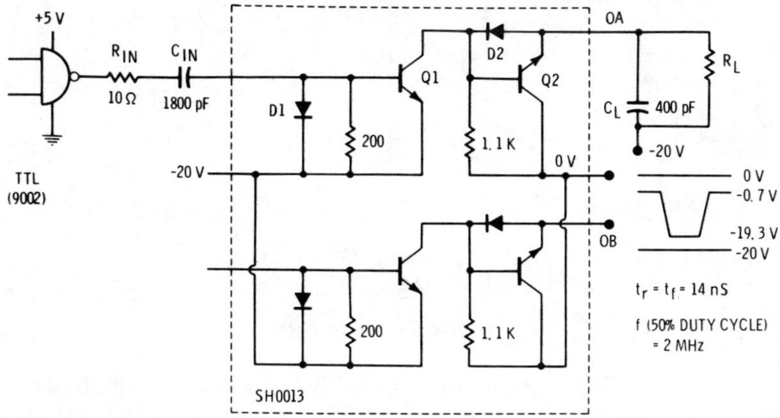

Fig. 5-25. Clock driver.

When the input signal is steady, transistor Q1 does not have any base-to-emitter voltage and thus is off. The output voltage will be near 0 V, and load capacitor CL will be charged to the supply voltage, shown as −20 V on the drawing. Assuming the circuit being driven consists of MOS inputs, R_L is very large and can be ignored.

When the TTL gate goes from 0 to 1, the input voltage to Q1 begins to rise. When it reaches about 0.6 V, Q1 begins to conduct. This action connects output OA to −20 V through D2 and Q1. Load capacitance C_L discharges through this path. Resistor R_{IN} and capacitor C_{IN} together determine how long OA stays near −20 V. For the values shown, the output pulse width will be about 200 ns. When the input signal finally decays, Q1 turns off. With the output at −20 V, Q2 is biased into conduction and turns on. Current, therefore, flows through Q2 and charges C_L. The output voltage moves toward 0 V. The circuit stabilizes when the drop across Q2 is about 0.7 V.

When the TTL driving gate goes low, Q1 is not affected since the input signal is the wrong polarity. Diode D1 acts as a protective clamp.

Up to 600 mA of load current can be drawn from each circuit, and pulse widths from about 100 ns to about 800 ns can be obtained. Other driver gates, such as TTL buffers and line drivers, will produce somewhat faster rise times (leading edge transition of the output waveform).

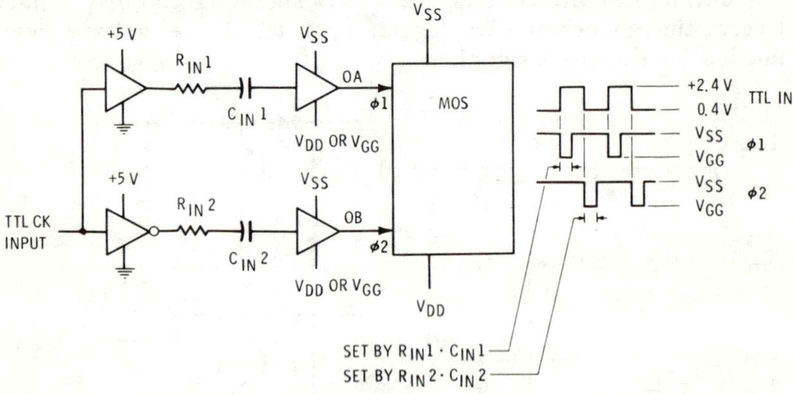

Fig. 5-26. Two-phase clock.

A 2-phase clock driver circuit can be constructed as shown in Fig. 5-26. The leading edge of the TTL clock generates Ø1, and the trailing edge generates Ø2. Pulse width of Ø1 is set by $R_{IN}1 \times C_{IN}1$, and pulse width of Ø2 is set by $R_{IN}2 \times C_{IN}2$.

6

NMOS

CHARACTERISTICS OF NMOS ICS

Conduction in n-channel MOS transistors is accomplished by means of electrons, whereas in p-channel devices it is by means of holes. Since electrons have been found to move faster than holes, n-channel devices are faster than p-channel. Recently introduced PMOS memory type SY1103A-X (by Synertek) has an access time of 110 ns, while equivalent NMOS memory type EA1500A (by Electronic Arrays) has an access time of about 60 ns.

The higher mobility of electrons over holes also allows NMOS transistors to be made smaller than PMOS transistors. This contributes to the higher speed of NMOS, and it also allows denser ICs—more transistors per square millimeter of silicon real estate. Ion implantation has become widely used in NMOS where it is used to lower the gate thresholds and to form depletion-type transistors that are used as load resistors. In certain circuits, a true resistor or nonsaturated load, rather than an MOS transistor load, allows the operating voltage to be reduced and therefore gives greater compatibility with TTL and other logic.

Another important characteristic of NMOS is operation on positive voltage. Many NMOS ICs are designed to work with a single +5-V power supply. Input and output logic levels in such cases are the same as TTL logic levels, and the interface between the two families is often directly compatible—NMOS outputs can drive TTL inputs directly, and TTL outputs can drive NMOS inputs directly without level shifting or special pull-up or pull-down circuits.

Some NMOS ICs, however, use more than one voltage, often +5 and −12 V.

The NMOS IC is still relatively new, and the major class of ICs so far developed is that of memories, since this class is probably the largest potential market for the technology. But NMOS is also used for other circuits, especially microprocessors, and is certain to be applied to most areas now served by PMOS.

Circuits and logic organization using NMOS are likely to be similar or identical to equivalent circuits in PMOS. The major differences will be voltage polarities and levels, operating speeds, and more complexity.

STATIC NMOS RAM

Perhaps the most widely used 1024-bit static memory is type 2102 and its many alternatives. The circuit is n-channel, silicon gate, and uses ion implantation to form depletion-mode load resistors and to lower gate thresholds to the same level as TTL gates. The device is available in various speeds from about 1000-ns access time to about 275-ns, and with power consumptions from 300 mW per package to 225 mW.

The memory is organized as shown in Fig. 6-1. Circuit operation is similar to previously discussed static memories. Five ad-

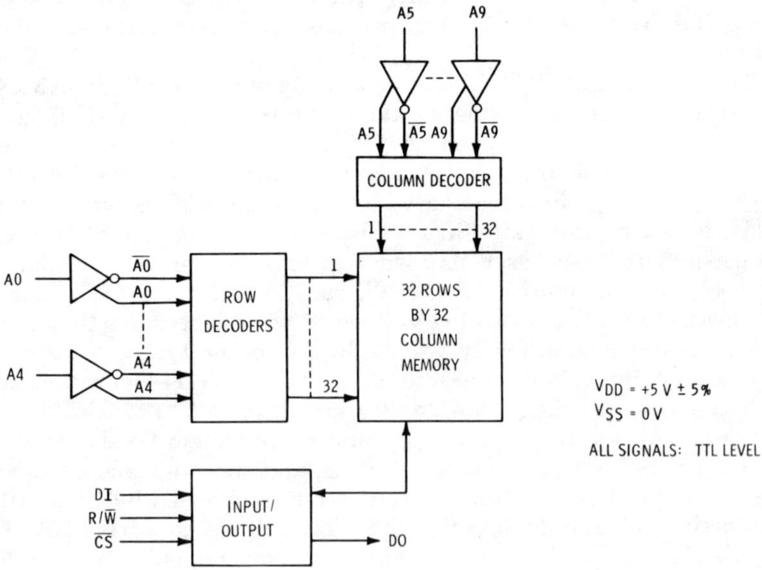

Fig. 6-1. Type 2102, 1024-bit, static NMOS memory.

dress bits select 1 of 32 rows, and five other bits select 1 of 32 columns. Then, either the bit at the intersection is read out or new data are written in. The chip-select input controls the 3-state output and the write circuit. When Chip Select is high, the output is a high Z, allowing OR-tie operation, and new data can be written in. The address bits are active, however, and data will appear at the output after the access time of the particular device.

The memory cell, shown in Fig. 6-2, is a cross-coupled flip-flop with two access transistors. If point M is high, point N is low. When the row is selected, Q_A and Q_B turn on, and line A follows point M and line B follows point N. When the column is selected, Q1 and Q2 are turned on and the differential output amplifier receives the two signals A and B, one high and the other low. If the chip is selected, the output goes high or low, depending on the data stored in the cell.

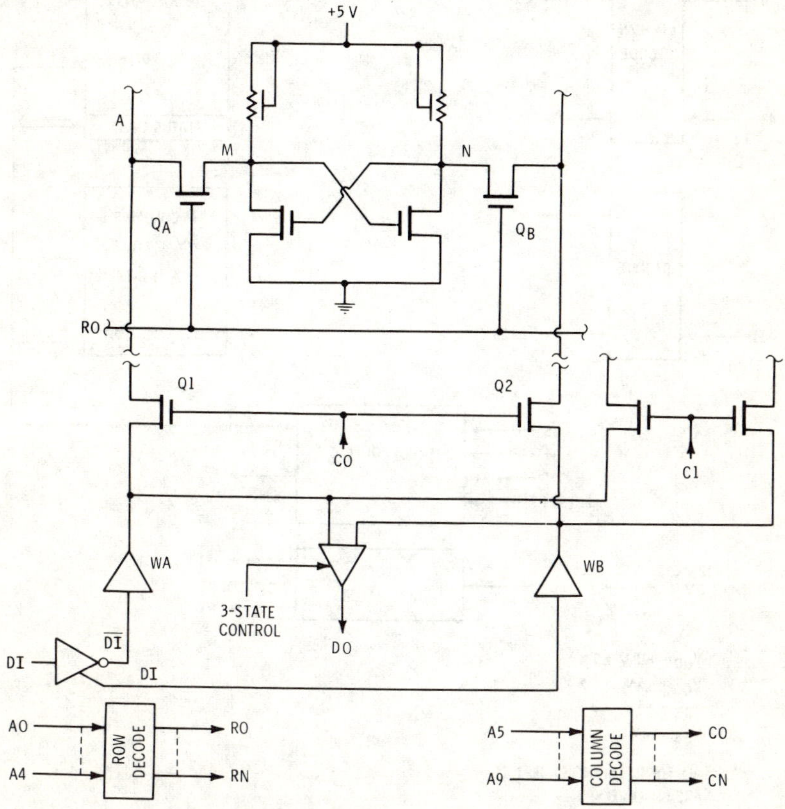

Fig. 6-2. Memory cell for NMOS static RAM.

To write in new data, the chip is selected; then the R/W input is brought low. The data at the DI input will then be written into the cell selected by the signals on the address inputs. When the circuit is in the write mode, the input data buffer generates signals D and $\overline{D}$, which are fed into the selected cell via write amplifiers WA and WB.

The IC is easy to use since it is fully compatible with TTL systems, using only one power-supply voltage, and can be OR-tied for expansion.

DYNAMIC 4K NMOS RAM

Inverting-type memory cells have been used extensively in dynamic NMOS RAMs, and these provide another option in the design of memory ICs. Fig. 6-3 shows the block diagram of a

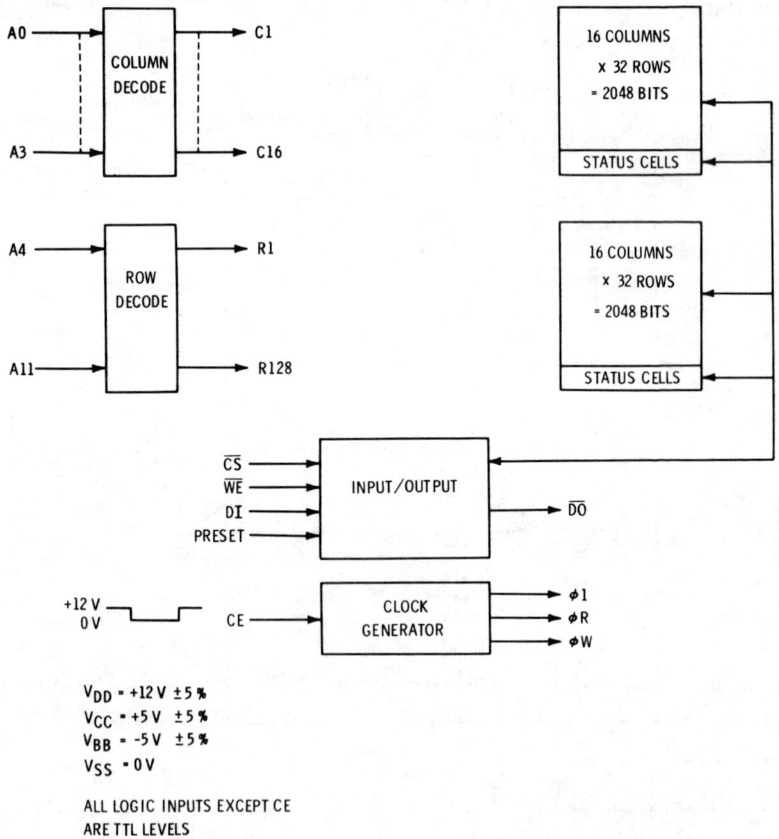

Fig. 6-3. Type RM1701, 4096-bit, dynamic NMOS RAM.

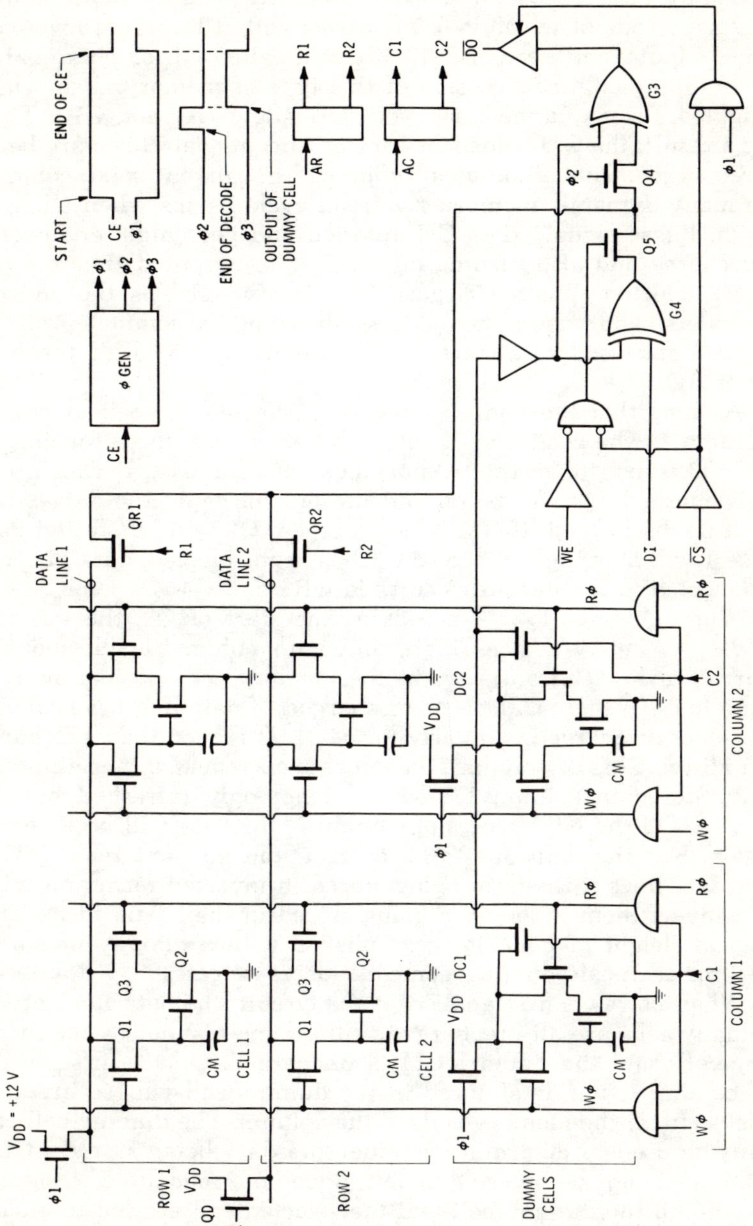

Fig. 6-4. Simplified 4-cell version of a 4k dynamic NMOS RAM.

typical 4k NMOS dynamic memory—in this case, type RM1701 made by Western Digital Corporation. The memory requires four voltages (one of which is 0 V), works with TTL signal levels for logic inputs, and needs a +12-V clock, called CE or chip enable.

A simplified 4-cell version of this type of memory is shown in Fig. 6-4. Prior to a memory cycle, CE is low (A) and Ø1 is high; as a result, the QD transistors are on and the data lines are being precharged. This eliminates the precharge current spike common to many dymamic memories. A read cycle begins when CE goes high. Phase signal Ø1 is CE inverted and goes high, ending the precharge and also turning on the 3-state output if the chip has been selected. When CE goes low, it also enables the address decoders, and when the address decoding is complete, this is sensed and used as a start for Ø2, which is also called the read Ø or RØ.

Assume that the address bits have selected the cell at row 1, column 1. The read line of cell 1 therefore goes high, turning on the Q3 transistor in the memory cells of column 1. If CM in cell 1 is charged, Q2 will be on and the precharge on the data-1 line will be discharged. If CM is not charged, Q2 will be off and data line 1 will stay high. Since row 1 has been selected, data line 1 is connected to the output circuit via QR1 and Q4.

When Ø2 goes low, it serves as the start of Ø3, the write Ø (WØ). When WØ goes high, and with column 1 selected, G2 turns on the Q1 transistors in the memory cells of column 1. If data line 1 is high, CM becomes charged; if data line 1 is low, CM becomes discharged; similarly for cell 2. It is seen that the charge on all the CMs of column 1 has now been reversed, and thus the data stored in column 1 have been not only refreshed but inverted. On the next access of column 1, the data will be inverted again. For the data out to be correct, the memory must invert the data lines if they are being stored in inverted form and must not invert them if they are being stored in their true form. This logical sleight of hand is accomplished with a dummy or status cell for each column, plus some exclusive-OR gates.

When power is first applied to the circuit, the data lines of the dummy cells are all precharged in the same manner as the memory cells and the dummy CM is uncharged. On the first access of column 1, the data stored in the dummy cell will be inverted just as in all the memory cells of the column. The dummy cell can thus be used to determine whether the data being stored in the column should be inverted or left alone on a readout. The signals from both the dummy cell and the selected cell are fed to exclusive-OR gate G3. If the signal from the dummy cell is high, the data from the memory cell are to be inverted. With the dummy-cell signal high and the selected data line high, the output of G3 will be low, but if the data line is low, G3 will be high. Thus, data inversion is obtained. If the signal from the dummy cell is low, G3 follows the data line without inversion.

Data are written into the cell by bringing $\overline{WE}$ low for about 160 ns while CS is low. Gate G4 will be following input DI and the output of the selected dummy cell in the same manner as G3. Thus, data are written into the selected cell either in true or inverted form to agree with the status of data already stored in the column by previous writings.

A number of similar NMOS RAMs have been developed, and a single-transistor cell of the type shown in Fig. 6-5 is used in many of these ICs. To read data out, the row-select line, RS, is brought high, and D/S line goes to the same state as CM and is sensed via a column-select signal, CO. If the data line goes high, the circuit acts to discharge CM; if D/S goes low, the circuit charges CM. Again, the data are refreshed but inverted. A dummy cell provides information on whether the data being stored is true or inverted, and the output and input circuits function in a similar manner to the 4k NMOS RAM discussed previously.

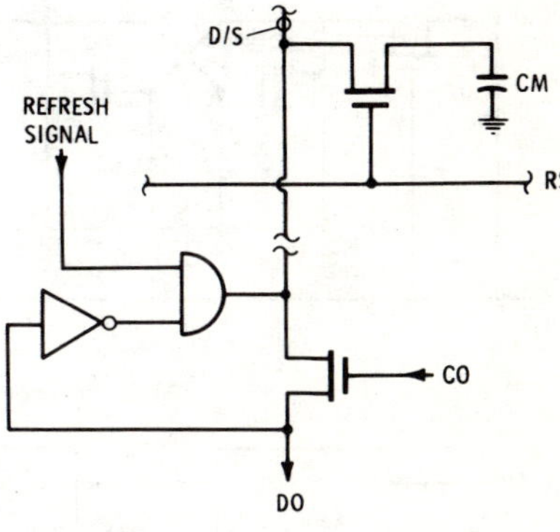

Fig. 6-5. One-transistor memory cell for NMOS dynamic memories.

QUASI-STATIC NMOS RAM

While dynamic memories offer high storage capacity at the expense of external refresh circuitry, static memories offer simpler operation at the expense of some storage capacity. In an effort to obtain the best of both worlds, the charge-pump memory cell was developed.

The cell, shown in Fig. 6-6A, uses the gate capacitance to store data, and since the charge will leak off, the cell must be refreshed and is therefore dynamic. Refreshing is obtained via the charge pumps, which are actually MOS transistors of which only two terminals are used. These two transistors are fed by an external oscillator operating continuously at about 1 MHz.

Assume C1 has been charged via access transistors Q_A and Q_B. Then, Q1 will be on while Q2 will be off and a high Z. Every time the pump voltage is +12 V, point B rises toward +12 and C1 is in the charging mode, while Q1 acts as a low Z across C2 and prevents charge buildup. When the pump voltage is at −8 V, the voltage at point A, which is near 0 V, and at point B, which is near V_{DD}, begins to charge toward −8 V. Because of the impedance of the charge-pump transistor, however, the voltage at A and B cannot change quickly; thus, Q1 remains on, preventing C2 from being charged negatively but allowing C1 to discharge slightly. On the next positive excursion, the charge on C1 is replenished, while C2 remains shunted by Q1. The pump

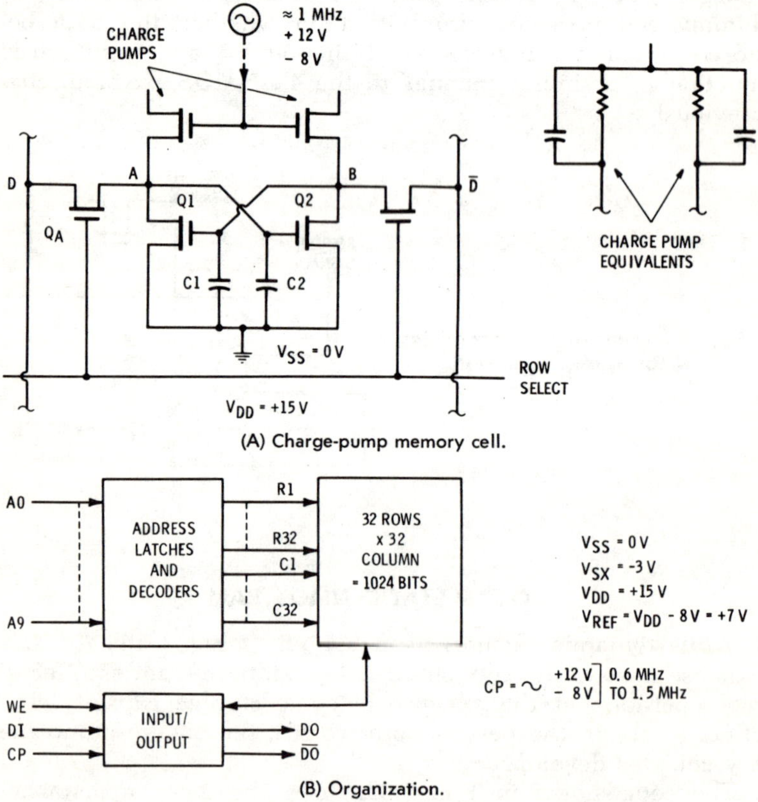

(A) Charge-pump memory cell.

(B) Organization.

Fig. 6-6. Type MW7001, 1024 bit, charge pump, static NMOS RAM.

voltage of +12 V and −8 V favors preservation of charge on the already charged C1.

The charge-pump cell is used in static-memory type MW7001, shown in Fig. 6-6B. Operation is typical of static memories having the added feature of address latches, which relax timing requirements, while data out is available, both true and complemented. The charge-pump voltage can be considered to be an ac bias signal rather than a refresh operation, since refresh is automatic.

16K NMOS ROM

The largest semiconductor memory chips available at this time are probably the 16,384-bit ROMs, although 32k-bit and 64k-bit chips will be along soon. Type RO-3-16384, manufactured by General Instrument Inc., is shown in Fig. 6-7. The memory is

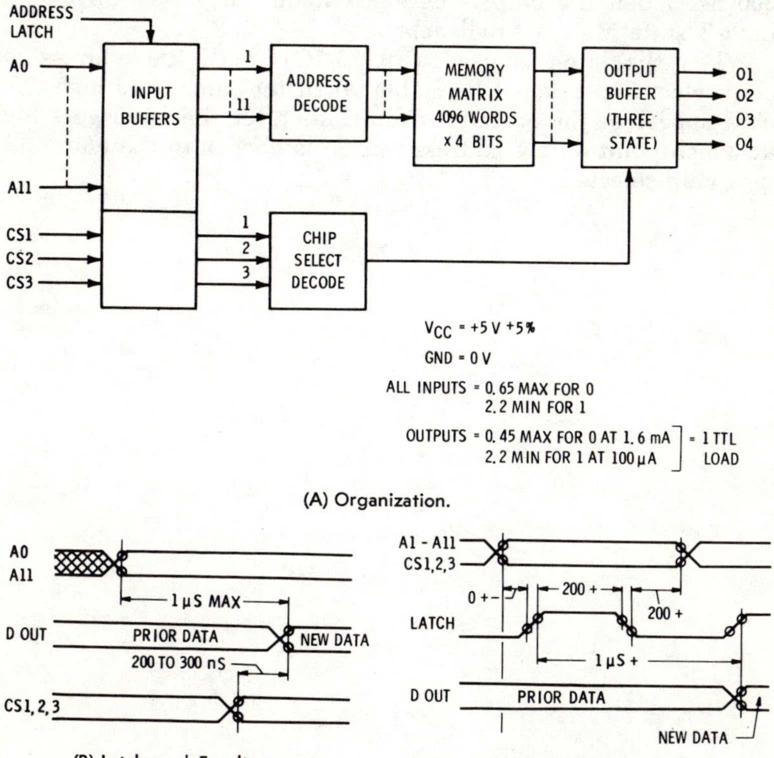

(A) Organization.

(B) Latch = +5 volts.

(C) Control by latch.

Fig. 6-7. Type RO-3-16384, 16k bit, NMOS RAM.

organized as 4096 words by 4 bits. It provides four outputs and requires 12 address bits. The internal structure of the ROM is straightforward, and its operation is indicated in Figs. 6-7B and C.

Operating features include an address-latch input and three chip-select inputs. The three chip-select inputs provide an 8-bit code that allows 8 ROMs to be operated as one 16k-×8-bit, or 128-bit, memory organized as 48k words of 4 bits each. The code for the chip-select input is specified by the customer and incorporated into the ROM in the final masking stage, just as the contents of the memory matrix are.

The ROM can be operated in two modes, either with the latch input high, as shown in Fig. 6-7B, or under control of the latch, as in Fig. 6-7C. If the chip is not selected, the outputs are in the high-Z state. For any given address, the output goes to the programmed state, once the chip is selected, with a delay of 200 to 300 ns. When the chip is deselected, the outputs return to the high-Z state after a small delay.

When the latch is used (Fig. 6-7C), both the address and chip-select inputs must be stable when the latch goes high. New data appear at the outputs within 1 μs after the latch goes high, at which point a new address can be latched into the chip and a new chip selected.

7

More MOS

COMPLEX MOS ICS

Probably the most important characteristic of MOS ICs is the great complexity that can be obtained on a small chip. This capability for complexity found its first major use in memory devices, primarily in computers, but also in communications systems and instrumentation. But the potential for many types of complex circuits exists, and special circuits are slowly being developed to fill specific needs. The larger the potential market for a certain circuit, the sooner a special IC is likely to be developed. Probably the largest class of complex circuits to be developed after memory devices has been calculator circuits. Next are digital watch and clock circuits, and then a host of others, with new ones appearing regularly. Space allows only a brief look at some typical devices in this class.

Calculator Circuits

These are the chips and chip sets that are used in pocket calculators and electronic slide rules. The number of chips and chip sets already developed is relatively large, and more chips and chip sets can be expected. The circuits perform all the regular mathematical functions, including logs, trigonometry, and exponentials. Because of the power of these circuits, they can be expected to show up in other applications, particularly control circuits. The block diagram for a typical low-cost pocket calculator is shown in Fig. 7-1.

IC Music

A number of ICs have been developed to aid in the generation of music. A complete electronic organ is relatively complicated,

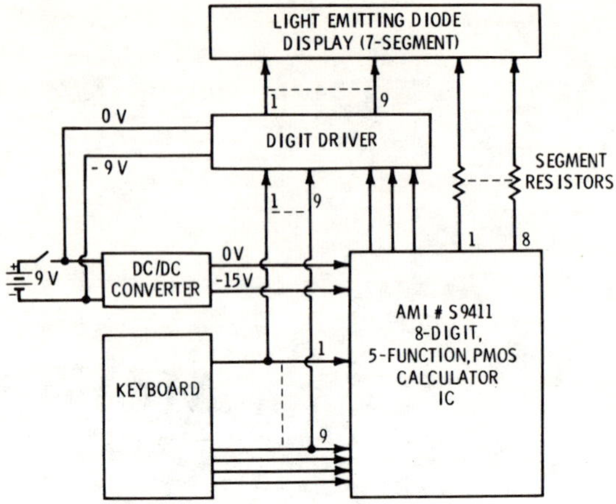

Fig. 7-1. Low-cost, portable calculator.

since in addition to audio signal generation and shaping, it requires a keyboard, foot pedals, power supplies, and an audio system, including amplifier and speaker. Much of the signal processing can be accomplished with relatively few ICs. American Microsystems, Inc., is an important supplier of PMOS ICs designed specifically for electronic music. Similar devices and some pin-for-pin equivalents are also available from other manufacturers.

Together, types S2555 and S2556, as pictured in Fig. 7-2A, produce the 13 highest frequencies used by an electronic organ when driven by an external oscillator. The output frequencies are not quite the same as those of the even-tempered scale but differ from it by a minimum of 0.112% for G♯8 to 1.119% for A♯8. Type S1857 is a more complex IC and combines the functions of S2555 and S2556 into one IC.

Each tone for the top-octave generator can be divided down to generate the tones for all the lower octaves, as shown in Fig. 7-2B. Twelve of the top tones are divided down in the same way, with a total of 12 divider circuits.

Each tone is made available at the keyboard and can be selected by the organ player. The selected tones are fed into formant filters that act to shape the signal, emphasizing certain harmonics and suppressing others. The object is to obtain waveforms that will produce sounds like a violin, piano, saxophone, flute, trumpet, etc.

In addition to requiring tones and voicing, an electronic organ needs various rhythms. These are created by a rhythm-generator IC such as an S2566. The IC is essentially a ROM in which typical rhythm patterns have been encoded. The user selects the desired rhythm pattern by means of switches; the tempo of the beat is controlled by a separate oscillator. A number of outputs are available whose waveforms are shaped by other formant filters to sound like bass drums, snare drums, cymbals, maracas, etc.

The waveforms of tones and the rhythm must then be combined in the audio system and fed to the speakers.

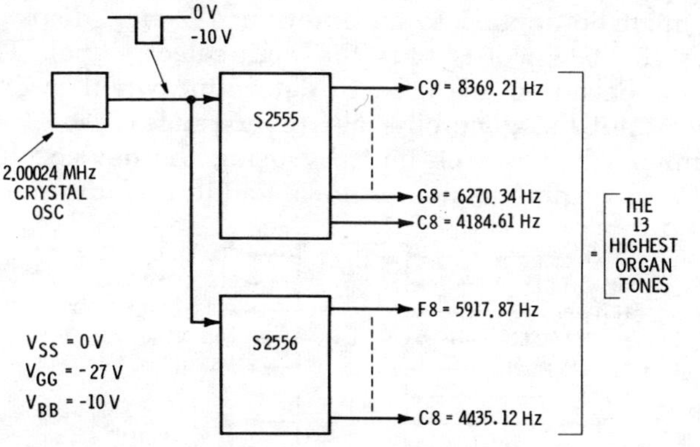

(A) Generator for the 12 top octave tones, plus C9.

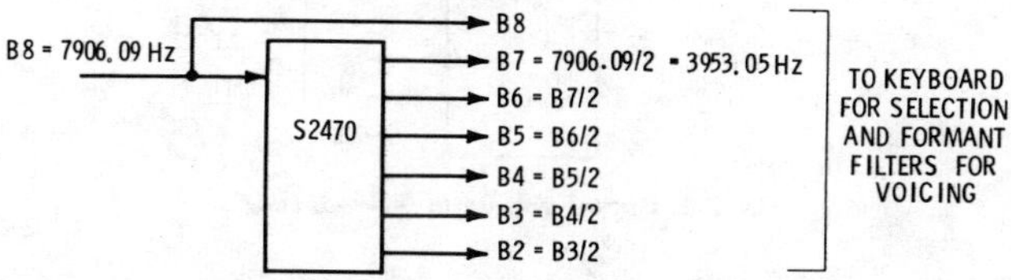

(B) Divider system for obtaining B tones.

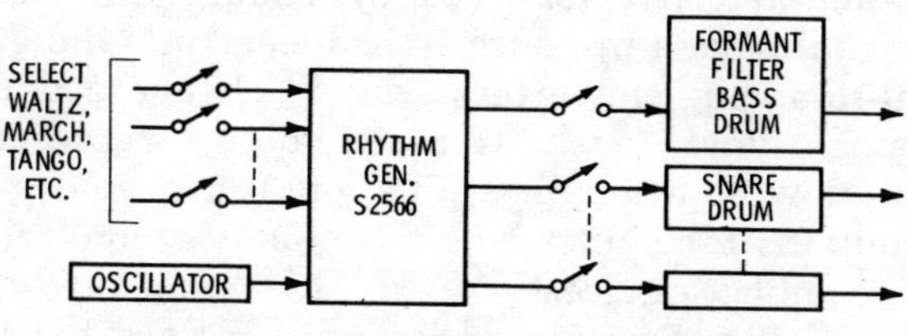

(C) Rhythm selection, generation, and voicing.

Fig. 7-2. Electronic music, using ICs.

Digital Voltmeter

The logic for building a 4½-decade digital voltmeter is contained in type 3814, shown in Fig. 7-3. The circuit uses dual-slope integration, which cancels the effects of long-term drift of components and voltages and rejects 60-Hz line noise. The clock can run at any frequency from dc up to 600 kHz, with a typical pulse width of 220 ns.

The unknown voltage first goes through an external conditioning circuit and then is applied at the Transfer input. The output consists of a multiplexed 4-bit bcd signal representing the numbers 0 through 9, plus a decoded digit-select signal. The digit-select outputs are under the control of the Step input. Thus, the outputs must be applied to an external 4½-digit display system that uses the bcd words plus the digit-select signal. The three outputs for decade 5 are used to determine whether or not the unknown input has gone off-scale (over-range).

A complete dvm can be built by using the device plus six or seven ICs, a display system, and a handful of discrete components.

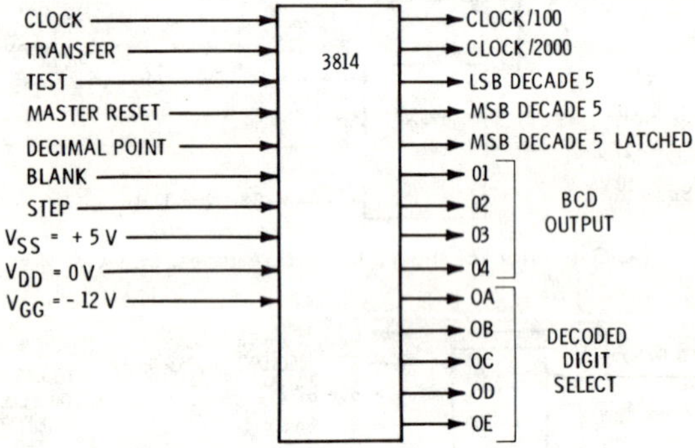

Fig. 7-3. Digital voltmeter—4½-decade.

Digital-to-Analog and Analog-to-Digital Converters

Analog-to-digital converters (adc's) change an analog input (such as a signal from a pressure transducer) into a digital word, while digital-to-analog converters (dac's) change a digital word into an analog signal (such as a dc voltage representing the pressure, the flow of water in a pipe, etc.). An analog-to-digital converter contains a complete digital-to-analog converter and is thus a more complex device.

Important specifications for adc's and dac's are the resolution or number of bits they can control or work with, fidelity of conversion (freedom from mistakes), speed of operation, and cost.

The dac type 3750 is shown in Fig. 7-4A. The two reference inputs provide high and low levels for full-scale and 0 outputs. Set and Reset are used to calibrate the circuit. The clock controls

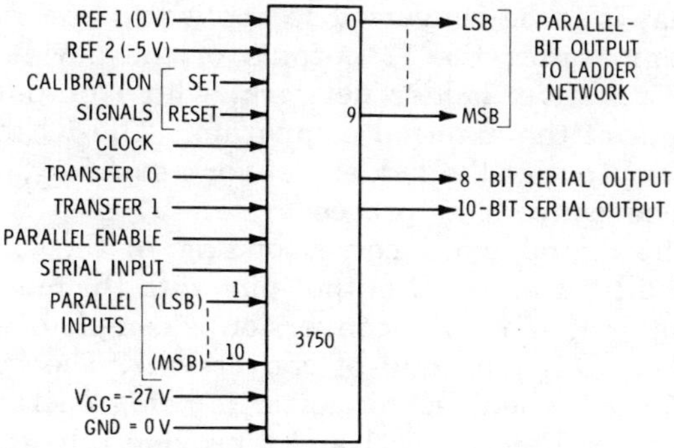

(A) Ten-bit dac.

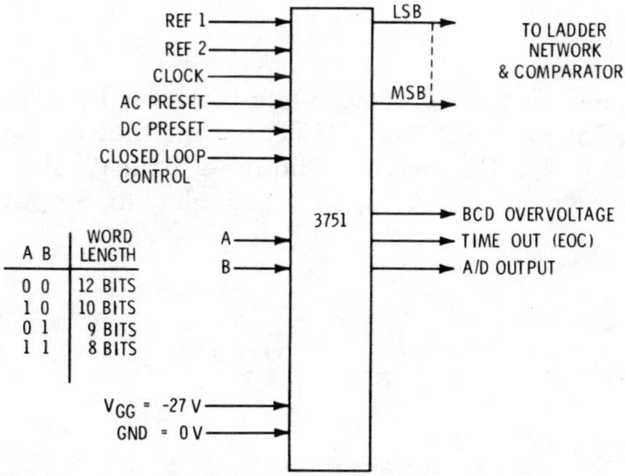

(B) Eight- to twelve-bit dac.

Fig. 7-4. Digital-to-analog and analog-to-digital converters.

the conversion rate. If data are entering serially, the clock can run at a maximum of 250 kHz; but if the data are being entered in parallel, the device can convert 500,000 digital words per second into analog signals. Ten inputs are provided for parallel entry and one input for serial entry.

Other inputs turn the device on and off and supply power. The ten parallel outputs are used to supply a ladder network of resistors in which the analog signal is created as a function of the outputs that have been turned on. The converted signal is also available in serial form as either 8 bits or 10 bits. This feature allows the data to be transmitted to a remote location over just two wires.

Fig. 7-4B shows the inputs and outputs of an 8-bit to 12-bit dac, with word length selectable by inputs A and B. The analog signal that is to be converted is applied to one input of the external comparator; the 12 outputs of the device are fed to an external, resistive, ladder network, whose output is fed to the other input of the external comparator. The IC then operates to make the output of the ladder network equal to the input analog signal. The conversion proceeds step by step with each clock pulse. The digital word corresponding to the analog input appears serially at the A/D output pin, with the most significant bit appearing first. When the conversion is complete, a pulse appears at the Time Out pin (end of conversion). The bcd Overvoltage output is used when the converter is being operated as a 3-digit bcd converter. References 1 and 2 provide full-scale and 0 references, and the other inputs are used to control circuit operation.

Wristwatch ICs

A number of manufacturers have developed ICs for wristwatches, such as type CT6002 manufactured by Cal-Tex Semiconductor, Inc. The IC is a CMOS circuit that is operated by a 1.5-V battery. The IC drives a liquid-crystal display (LCD) of the type shown in Fig. 7-5A and provides an output signal for

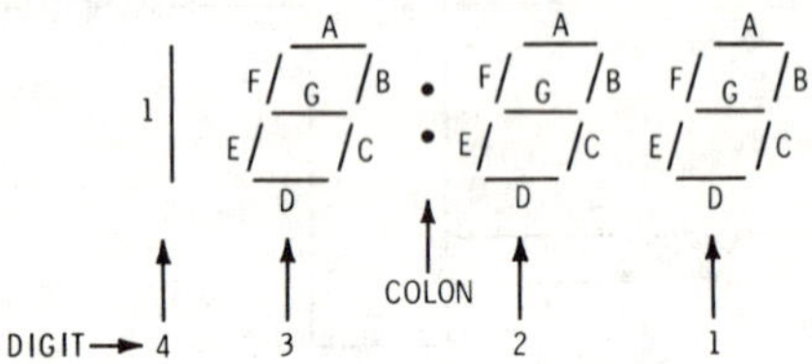

(A) Display arrangement.

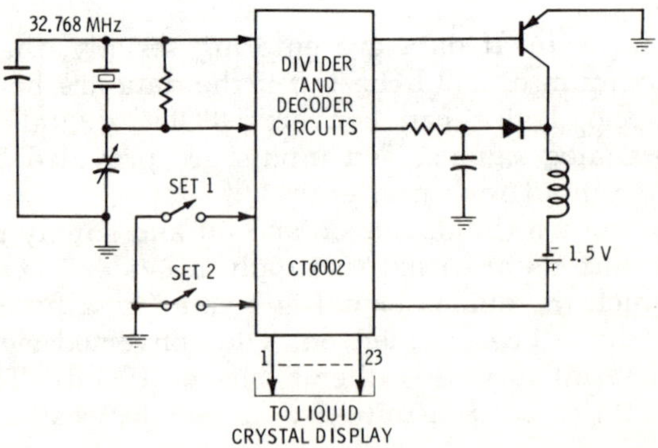

(B) Watch logic.

Fig. 7-5. Digital wristwatch.

each segment of the four figures, plus the colon. The display thus requires 23 connections from the IC.

The logic of the IC is shown in Fig. 7-5B and consists essentially of a digital counter, plus a decoding circuit to generate the 22 outputs, plus one common backplane driver. The external circuit consists of a 32.768-MHz quartz-crystal oscillator circuit plus a power converter to generate the eight volts needed by the LCD. Since the display elements need ac, the display signals from the IC are pulsed at 32 Hz.

Somewhat more complicated but logically similar circuits are available for operating digital alarm clocks and calendar watches and clocks.

Universal Asynchronous Receiver/Transmitter (UART)

Many data-handling machines, such as point-of-sale terminals, teletypewriters, display systems, etc., must communicate regularly with other remotely located pieces of equipment, such as a central computer (see Fig. 7-6). In many cases, the circuits needed to handle the signals at these interfaces between the equipment and the data link are very similar, and thus special ICs called universal asynchronous receiver/transmitters have been developed to do the job.

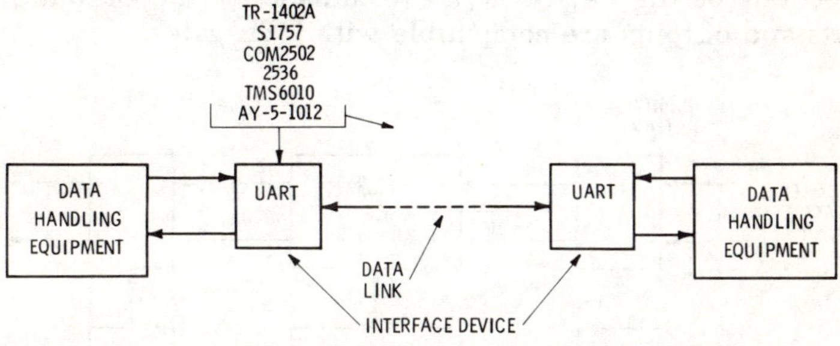

Fig. 7-6. Universal asynchronous receiver/transmitter (UART).

The UART is relatively complicated and performs all the transfer operations between equipment and line. The device can accept parallel data from the equipment and convert it to serial form for transmission over the data link. It also accepts serial data from the line and converts it to parallel form, if necessary, for use by the equipment. It also acts to add start, stop, and parity bits and to verify these bits in received messages.

Data can be handled in variable word lengths of 5 to 8 bits. Signal inputs and outputs are at TTL logic levels, and power supplies are +5 V and −12 V.

First-In, First-Out (FIFO) Register

In some applications it is desirable to accumulate data at variable rates and then deliver all the stored data in one quick dump cycle to a communications link that is perhaps feeding a computer. The reverse situation also occurs, in which a cluster of data is accepted and then fed out slowly to a peripheral device such as a printer.

The above operation can be accomplished with shift registers of suitable length except that data placed in the register are not immediately available at the output but must work their way through as new data are entered. What is needed is a circuit that accepts data and immediately shifts them through toward the output. The new data must not push out already stored data but should simply line up behind them like cars at a red light. Signetics type 2535 (organized as 32 words of 8 bits each) and Fairchild Semiconductor type 3341 (organized as 64 words of 4 bits each) are FIFOs of this kind (Fig. 7-7). Data enter at the command of the input control logic and are moved into the RAM and toward the output stage, piling up behind any other data previously stored. The control logic for the RAM keeps track of how many words have entered the register and how many have left, and controls the RAM accordingly. Data are moved out of the device at the command of the output logic. Inputs and outputs are compatible with TTL gates.

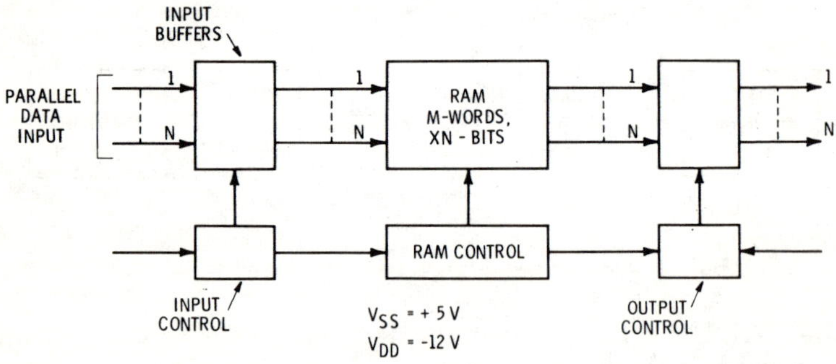

TYPE	N	M
2535	8	32
3341	4	64

Fig. 7-7. First-in, first-out register (FIFO).

Programmed Logic Arrays

Something of a cross between a ROM and a complex block of logic is the programmed logic array, or PLA. The basic idea of a

PLA is indicated in Fig. 7-8. First, a standard array of gates and flip-flops is designed on the chip. The chip has inputs A, B, . . . , N (usually from 8 to 14) and 8 or more outputs. The outputs are various functions of the inputs as indicated (where X and Y are functions of inputs not shown, and M and N are ANDed functions of certain inputs). The lines with circles indicate connections that can be removed during the final masking stage—thus the resemblance to a ROM.

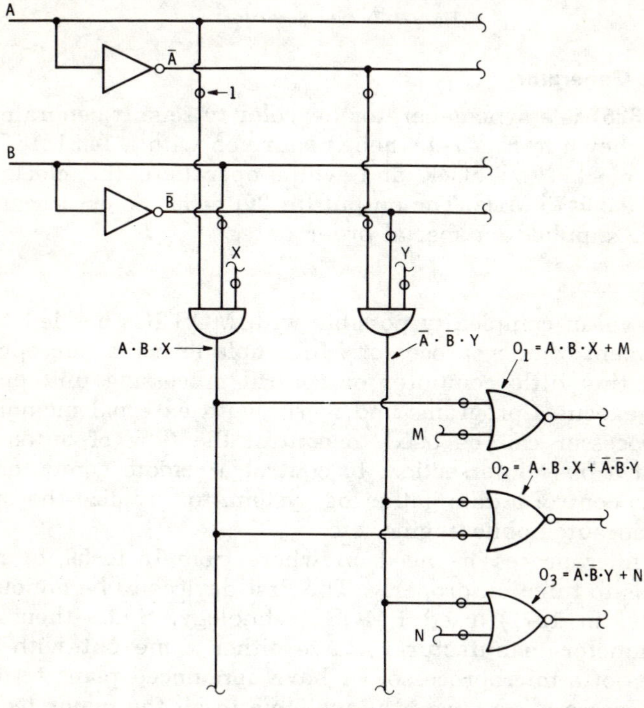

Fig. 7-8. Programmed logic array.

To make output O1 represent a different function than that shown, connection 1 can be eliminated, giving O1 = BX + M. The same is true for all the other outputs. In addition, feedback loops can be included such that O2 can be true only if O1 was true in the preceding clock cycle.

Relatively complicated logic can be accomplished with PLAs. The circuit designer writes the logic equations, and these are then usually fed into a computer, which determines which paths will be used in the PLA; the computer also generates the final mask.

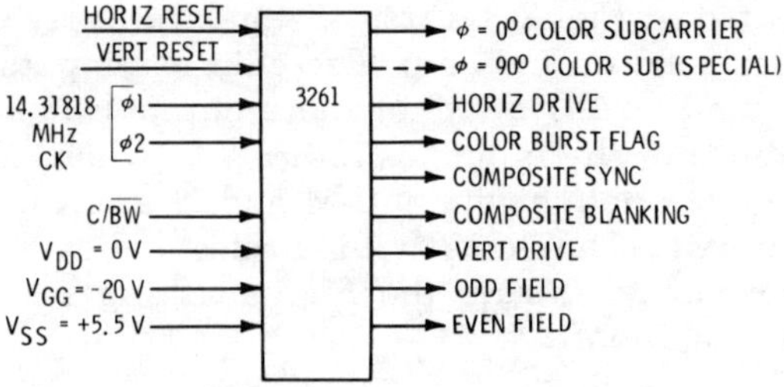

Fig. 7-9. Tv sync generator.

TV Sync Generator

Type 3261 is a sync generator for color-tv sets. It generates the outputs shown in Fig. 7-9 when it is driven with a 14.31818-MHz, 2-phase clock. For black and white operation, the clock rate should be 2.0475 MHz. The output for "Ø = 90° color subcarrier" output is supplied on special order only.

Microprocessors

The level of complexity possible with MOS ICs has led to the development of microprocessors. In simple terms, a microprocessor is a tiny little computer or central processing unit on one chip. It executes programs and works with external memory. A microprocessor can be used to control the flow of automobile traffic at a busy intersection, to control a remote computer terminal, to control a data-gathering system, to optimize the operation of an automobile engine, etc.

Most microprocessors need anywhere from 15 to 50, or more, other ICs to function properly. The first devices to be announced (by Intel in 1971) used PMOS technology. Since then, most semiconductor manufacturers have either come out with their own idea of a microprocessor or have announced plans to do so. Microprocessors are currently available in all the major technologies—PMOS, NMOS, CMOS, bipolar—and with a wide range of capabilities and features. Many more such devices will be developed in the future.

LOGIC INTERFACES

A large number of interfaces are possible between the three basic types of MOS and their many varieties and all the bipolar logic families. In all cases, a number of parameters must be compatible before two different families can run together. These parameters include voltage levels and polarities, input and output currents, impedances, waveform rise and fall times, protection against destructive transients, and system timing requirements.

Some interfaces between logic families are completely compatible, while most others can be accomplished relatively easily, although extra components such as pull-up and pull-down resistors are often needed. Sometimes level-shifting circuits are needed, and these as well as pull-up and pull-down circuits are sometimes built into the IC. Input requirements and output capabilities of many logic families are listed in Table 7-1.

The most common MOS interface is that between devices in the same family, such as CMOS to CMOS, as shown in Fig. 7-10A. The output voltage levels of 0.01 V for 0 and 4.99 V for 1 are more than one volt better than what the inputs require. Quiescent current levels are near 0. Fan-out to other CMOS inputs can be 50 or more and is limited by the speed at which the system operates. Maximum speed for CMOS gates operating at +5 V is about 5 MHz. With $V_{DD} = +10$ V, operating speed can be increased to about 10 MHz. With this value of V_{DD}, the input and output voltages are higher, and a gate operating at +5 V cannot drive a gate operating at +10 V.

In Fig. 7-10B CMOS gates are driving low-power TTL (LPTTL) gates. Voltage levels and currents are compatible, and 54C/74C can drive two LPTTL inputs, while 4000 devices can drive one.

The guaranteed output level for LPTTL is not high enough to drive a CMOS input directly (Fig. 7-10C), but a pull-up resistor raises the output level to above the 3.5 needed for logic 1. Since little current flows, the fan-out can be high.

A CMOS gate cannot drive a regular TTL input (see Fig. 7-10D), since it cannot sink the required 1.6 mA without the output rising above the needed 0.8 V. A CMOS buffer, however, such as 4049, can drive two regular TTL inputs. Also, the output drive of CMOS can be increased by operating one or more gates in parallel.

As shown in Fig. 7-10E, regular TTL gates cannot drive CMOS gates without a pull-up resistor. However, 8800 series TTL gates can (Fig. 7-10F), since they do not have a series diode in the output-high circuit and thus can go to 3.6 V, which is just above the 3.5-V minimum the CMOS gate needs.

High-threshold logic (HTL) and high-noise-immunity logic are used in many industrial applications because of their high resistance to noise. When a CMOS gate is operated at +15 V, as in Fig. 7-10G, it can supply one HTL input directly. The output of the HTL gate swings from 0.5 V to +13.5 V in normal operation, which is enough to drive the CMOS gate. However, the

Table 7-1. Major DC Characteristics of Digtal Logic Families

Device		Required Logic Inputs		Logic Outputs		in volts			
		Logic Low V_{max} (volts) @ I (mA)	Logic High V_{min} (volts) @ I (μA)	Logic Low V_{max} (volts) @ I (mA)	Logic High V_{min} (volts) @ I (μA)	V_{CC}	Gnd	V_{SS}	V_{DD}
TTL	Std.	0.8/1.6	2.0/40	0.4/16	2.4/400	5	0		
	H	0.8/2	2.0/50	0.4/20	2.4/500	5	0		
	LPTTL	0.7/0.18	2.0/10	0.3/2	2.4/200	5	0		
	S	0.8/2	2.0/50	0.5/20	2.5/1000	5	0		
	LS	0.7/0.36	2.0/20	0.5/8.0	2.7/400	5	0		
	8800	0.8/1.6	2.0/40	0.4/16	3.6/10	5	0		
High Threshold		6.5/1.2	8.5/2	1.5/12	12.5/170	15	0		
HNIL		5/2.6	6.5/10	1.8/13	13.0/10 mA	15	0		
ECL[1]		—1.5 V	—0.7 V	—1.55 V	—0.75 V				

CMOS[2]	4000	+5 V	1.5 V	3.5 V	0.4/1.6	2.5/3 mA			0	5
		+10 V	3.0 V	7.0 V	0.5/3.5	9.5/2.3 mA			0	10
		+15 V								
	54C/74C	+5 V	1.5 V	3.5 V	0.4/0.36	2.4/100			0	5
		+10 V	2.0 V	8.0 V	1.0/10 μA	9/10			0	10
		+15 V							0	15
	14500	+5 V	1.5 V	3.5 V	0.4/0.8	2.5/1700			0	5
		+10 V	3.0 V	7.0 V	0.5/2.0	9.5/900			0	10
		+15 V			1.5/7.8	13.5/3500			0	15
PMOS[2, 3]	Low Threshold		0.8 V	4.0 V	0.4 V	4.4 V			5	−12
	High Threshold		−9.0 V	−2.0 V	−16 V	0 V		0		−27
NMOS[2, 3]			0.6 V	2.2 V	0.4 V	2.4 V	5		0	+12

[1] $V_{CC} = 0$; $V_{EE} = -5.2$ V
[2] All MOS input currents typically 10 picoamperes.
[3] Much variation, see data sheets

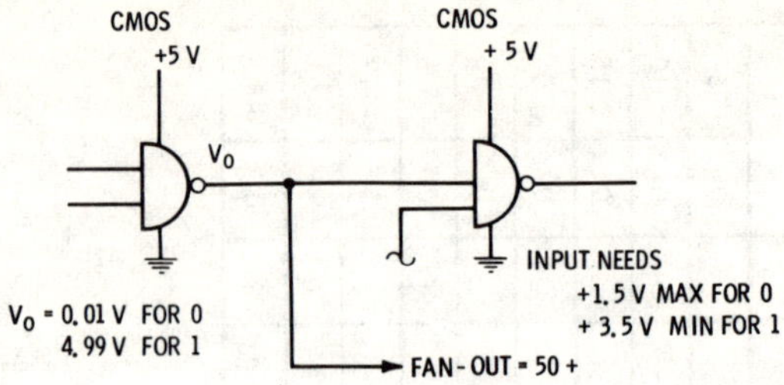

(A) CMOS to CMOS.

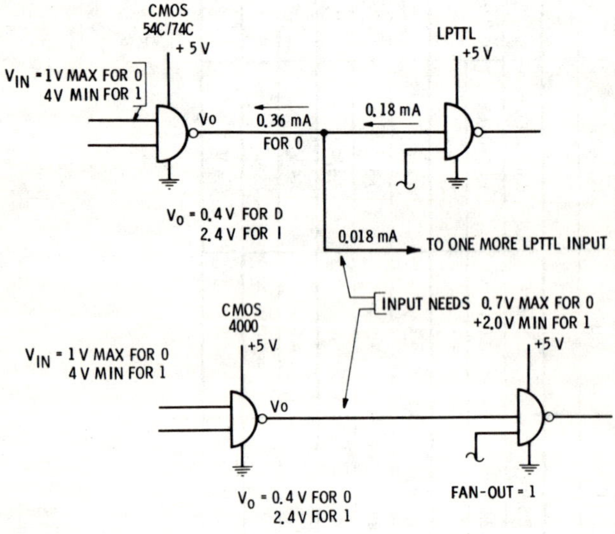

(B) CMOS to LPTTL.

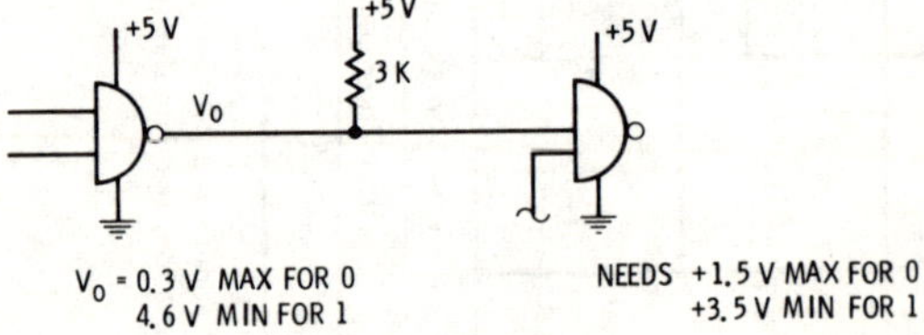

(C) LPTTL to CMOS.

Fig. 7-10. Common

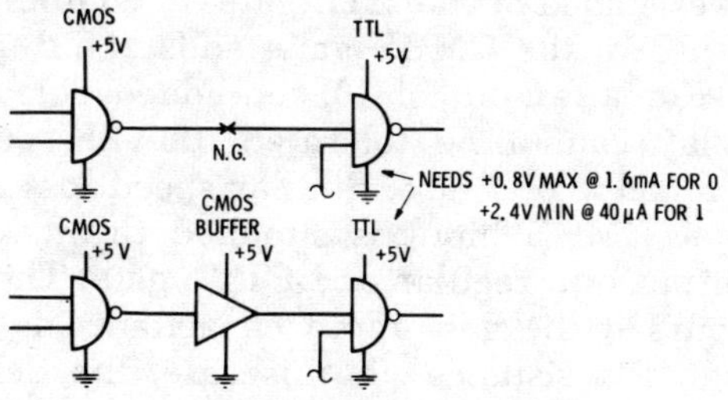

(D) CMOS TO TTL.

(E) TTL to CMOS.

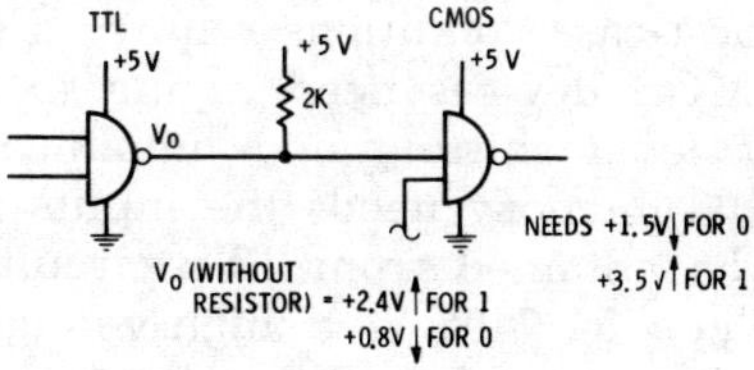

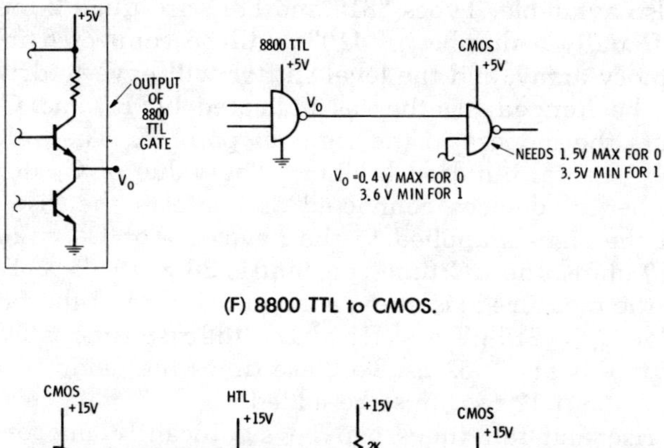

(F) 8800 TTL to CMOS.

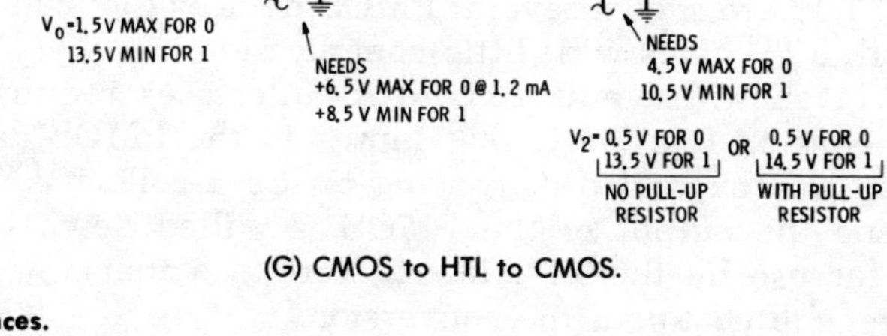

(G) CMOS to HTL to CMOS.

logic interfaces.

internal load resistance of the HTL gate, in combination with the input capacitance of the CMOS gate, adds about 75 ns for each CMOS input. For a fan-out of 10, the added delay is about 750 ns. If this delay cannot be tolerated, the 2K pull-up resistor reduces it by a factor of 7 or 8. Similar speed losses occur whenever a MOS or CMOS input is supplied through a resistance, such as the output of a regular or LPTTL gate. The same type of speed loss occurs if the gate input capacitance is discharged to ground through a resistance. In this case, the delay occurs on gate turn-off rather than on turn-on.

Fig. 7-11 shows a type 8800 TTL driving a 2500 series PMOS. If $V_{CC}1$ is tied to $V_{CC}2$ (Fig. 7-11A), the TTL gate can drive the PMOS input directly. If the two V_{CC}s are not tied together (Fig. 7-11B), worst-case conditions require a pull-up resistor.

Some PMOS devices need inputs so far removed from TTL levels that level shifting and amplification are required. The popular 1103 memory needs the inputs shown in Fig. 7-12, and these can be obtained from TTL circuits with the level shifter shown. (Type 54/7406 is a high-voltage—it will withstand 30 volts—hex inverter with open collector. A number of other TTL ICs with open-collector outputs rated to +14 V, +15 V, and +18 V are also available. Types 8810 and 8811 are quad 2-input NAND gates.) Usually a number of 1103s will be connected in parallel in a memory array, and the level shifter will have to drive a load that can be lumped together as indicated by R_L and C_L, which represents the devices to the right of point X. The value of R_L is usually several hundred kilohms. The value of C_L depends on the number of devices connected and affects the rise and fall times of the signals applied to the devices. For $C_L = 26$ pF and $R1 = 470$ ohms, the RC time constant is $26 \times 10^{-12} \times 470 = 12.2 \times 10^{-9}$, the measured rise time is about 34 ns, and the fall time is 19 ns. For $C_L = 215$ pF and $R1 = 1K$, the rise time is 500 ns and the fall time is about 52 ns. To these times the delay through the inverter (about 12 ns) must be added.

Both rise and fall times can be significantly improved if an active instead of a passive pull-up circuit is used: with $C_L = 200$ pF and $R1 = 1K$, the rise time is 30 ns and the fall time is 50 ns.

In Fig. 7-13 are shown several PMOS devices that can be interfaced with TTL gates with little complication.

Fig. 7-14 shows several TTL/MOS interfaces requiring level shifting. Note in Fig. 7-14C that inputs to the PMOS device are pulled up and expanded (by using an open-collector TTL output) while the output of the PMOS is pulled down and compressed for use by the TTL input. Note also that V_{SS} and V_{DD} have been shifted to aid the conversion.

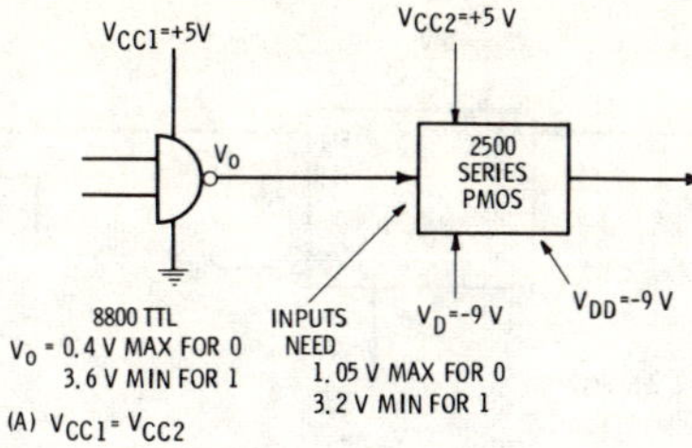

(A) $V_{CC1} = V_{CC2}$.

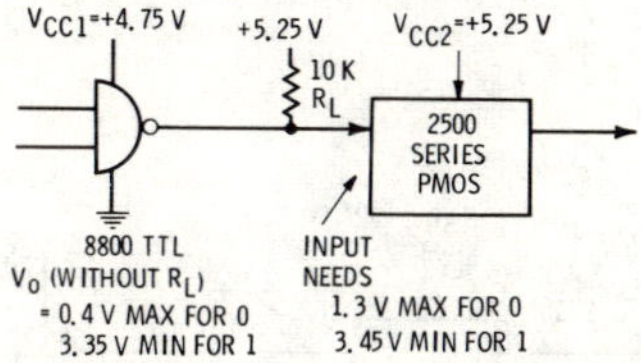

(B) Worst-case V_{CC1} and V_{CC2}.

Fig. 7-11. Interfacing 8800 TTL to 2500 PMOS.

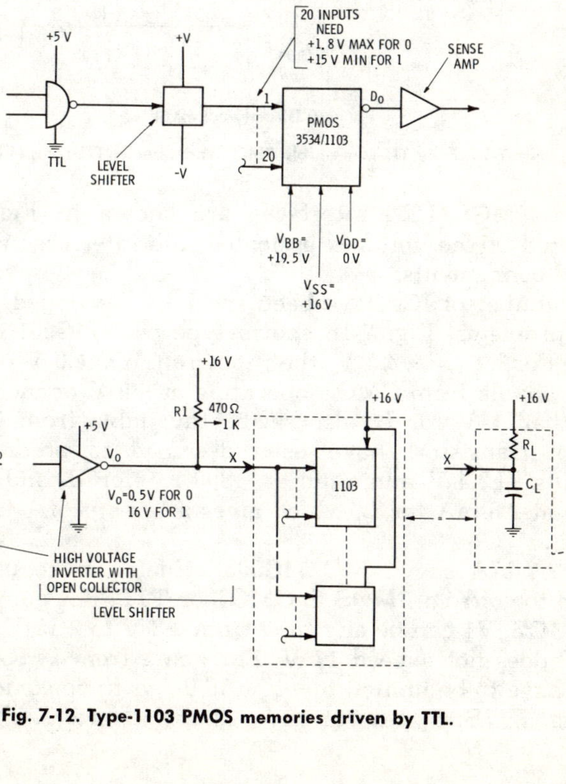

Fig. 7-12. Type-1103 PMOS memories driven by TTL.

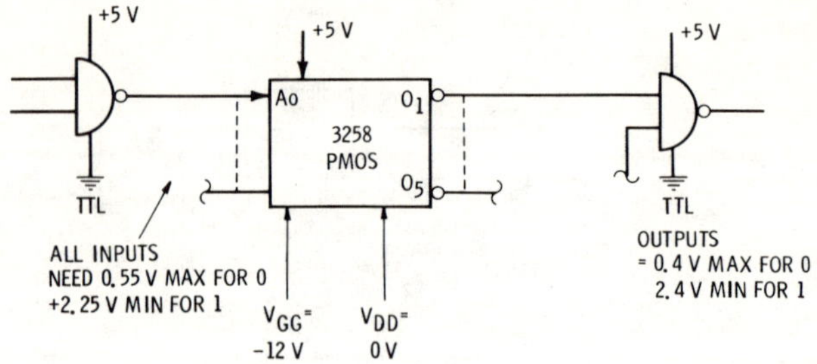

(A) Fully compatible.

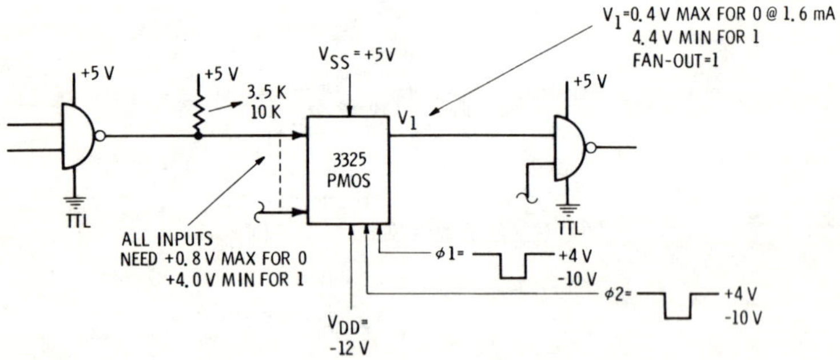

(B) Directly compatible.

Fig. 7-13. Fully TTL-compatible PMOS and directly TTL-compatible PMOS.

The NMOS/TTL interfaces are shown in Fig. 7-15. When operated at the voltages indicated, the interfaces require no external components.

A number of ICs have been specially developed to solve interface problems. Fig. 7-16 shows type 54C/74C901, for CMOS to TTL. For $V_{CC} = +4.5$ V, the input requires 4.0 V or more (readily available from CMOS operating at +5 V or more) for H, and less than 1 V for L, also readily available from CMOS. The p and n transistors have been designed to produce TTL-rated outputs: +2.4-V minimum at −800 μA for H, and +0.4-V maximum at 2.6 mA for L, which gives a fan-out of 1 to regular TTL inputs.

Fig. 7-17A shows 54C/74C903, which can be used as in Fig. 7-17B to go from PMOS to CMOS or TTL. For going from PMOS to CMOS, V1 can be anything from +3 V to +15 V, as long as V1 + V2 does not exceed 17 V. For going from PMOS to TTL, V1 will have to be limited to +5 V, if V_{CC} is to be common, as shown in Fig. 7-17B. Types 901 and 903 are inverters. Type 902 is the

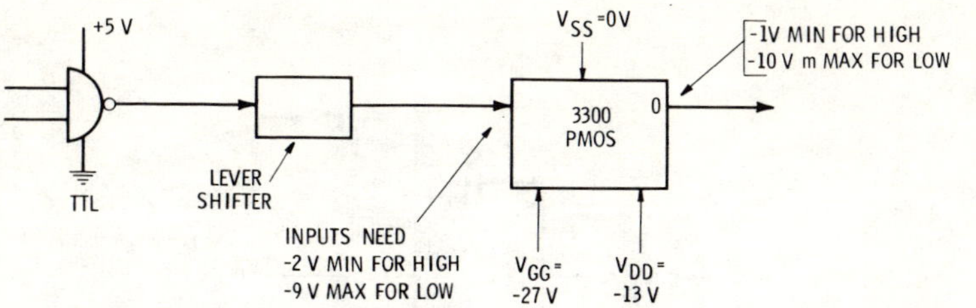

(A) Typical TTL to PMOS interface.

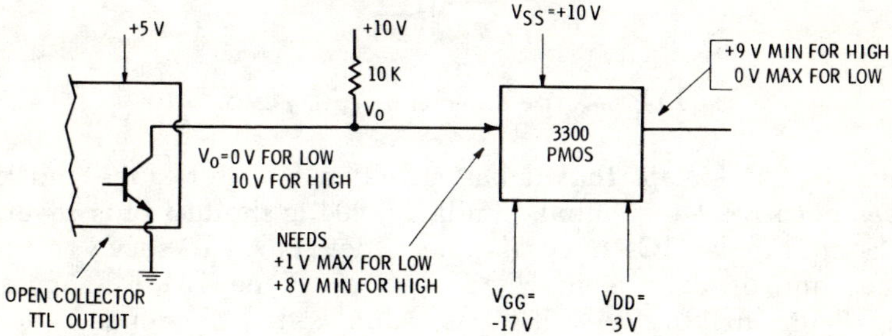

(B) Pulling up TTL outputs.

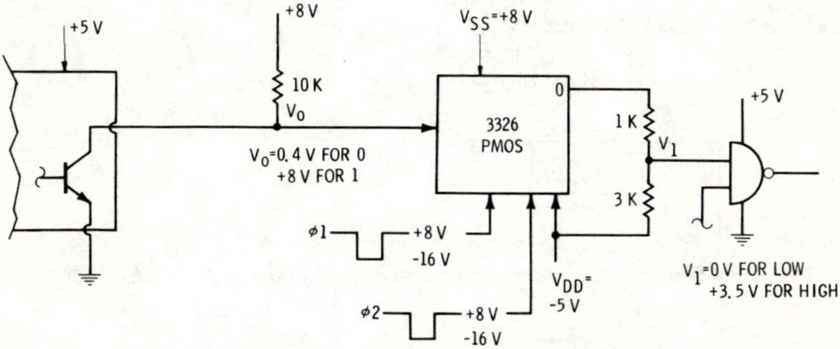

(C) Pulling up TTL and pulling down PMOS.

Fig. 7-14. Interfacing TTL to PMOS to TTL.

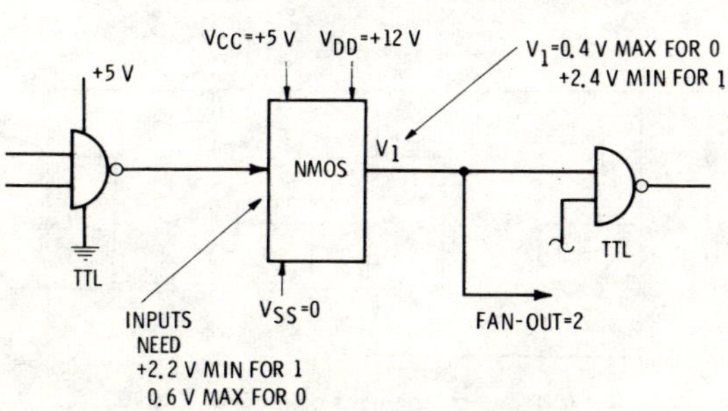

Fig. 7-15. Interfacing TTL to NMOS to TTL.

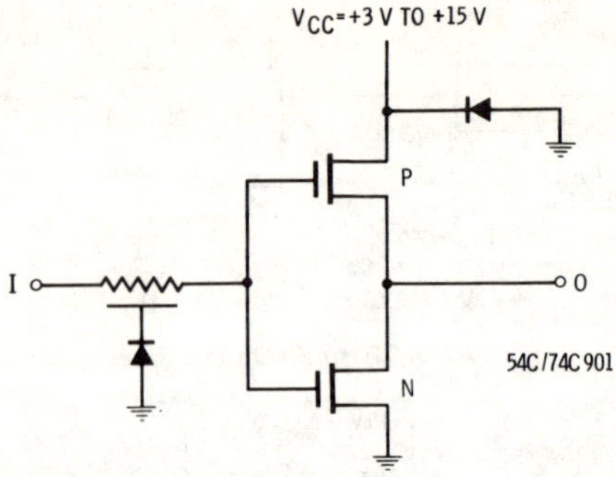

Fig. 7-16. Integrated circuit for interfacing CMOS to TTL.

same as 901 except that it has an extra inverter stage and thus does not invert the signal; similarly, 904 is the 903 plus inverter.

Some bipolar ICs have also been developed to serve as interfaces and/or drivers in MOS systems. Type 75451 (shown in Fig. 7-18A driving type 0027) is a dual-NAND, plus open-collector transistor. Each transistor can sink 300 mA and withstand 30 V. Similar devices, but providing AND and NOR gating, are types 75452 and 75453, respectively.

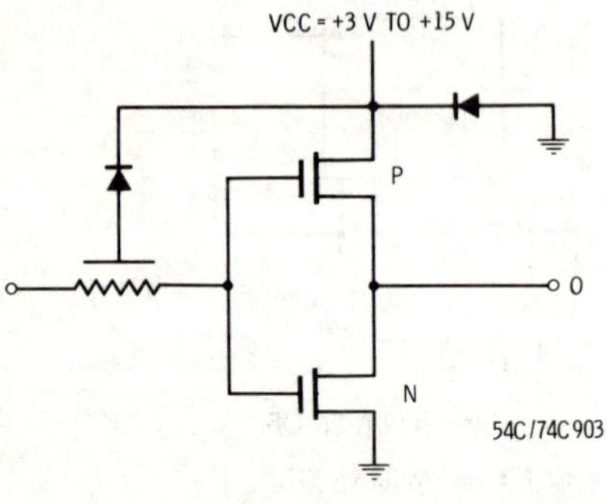

(A) Circuit using 54C/74C903.

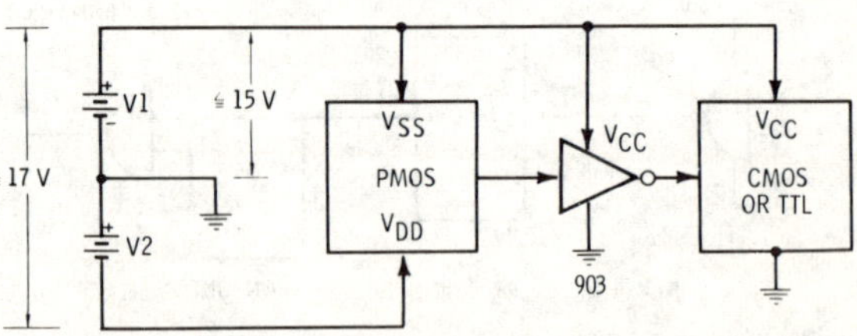

(B) Voltage arrangement using 54C/74C903.

Fig. 7-17. Integrated circuit for interfacing PMOS to CMOS or TTL.

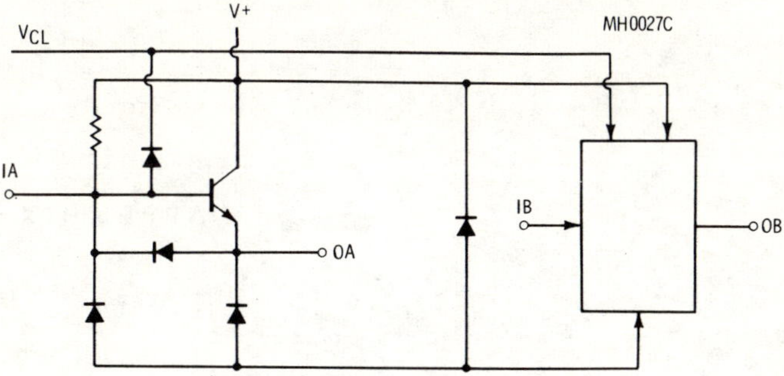

(A) Type 75451 driving type MH0027C.

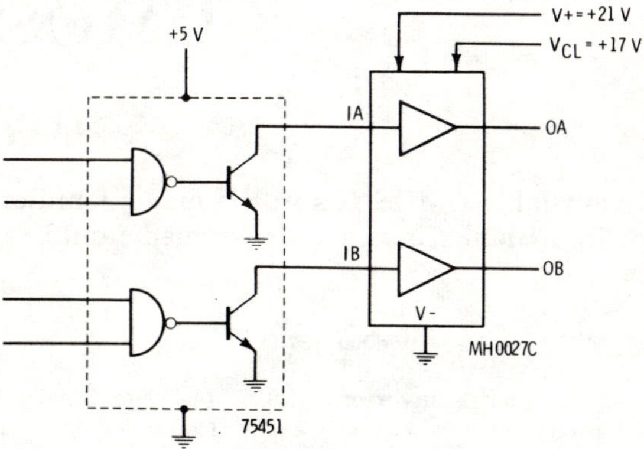

(B) Dual-interface driver.

Fig. 7-18. High-current, high-speed driver.

Type MH0027C, shown in Fig. 7-18B, is a dual high-speed interface driver that can supply intermittent peak currents of 1.2 amperes (each circuit) at 25 volts. The circuit is essentially a transistor surrounded by protective clamping diodes. Protection is necessary because the circuit can supply 1-A currents to 600-pF loads in 25 ns. Even small lead inductances can cause substantial overshoots and undershoots when such large currents are switched in such short periods. The circuit is designed primarily to supply MOS circuits, such as a group of 1103 dynamic memories. Note that the circuit is a source of current only, not a sink. To discharge the input capacitance of a MOS memory, a separate sink circuit such as an open-collector TTL gate can be used and tied to OA. The logic should then act to turn on either the source or the sink circuit, but not both.

APPENDIX **A**

CMOS ICs

Currently available CMOS ICs in the major families are listed in the following. Blanks indicate unassigned numbers or blocks of numbers.

4000 SERIES

4000	Dual 3-Input NOR + Inverter
4001	Quad 2-Input NOR
4002	Dual 4-Input NOR
4003	
4004	
4005	
4006	18-Stage Shift Register
4007	Dual Complementary Pair + Inverter
4008	4-Bit Full Adder
4009	Hex Buffer/Converter
4010	Hex Buffer/Converter
4011	Quad 2-Input NAND
4012	Dual 4-Input NAND
4013	Dual D Flip-Flop
4014	8-Stage Static Shift Register
4015	Dual 4-Stage Static Shift Register
4016	Quad Bilateral Switch
4017	Decade Counter
4018	Preset/Divide-by-N Counter
4019	Quad AND/OR Select
4020	14-Stage Binary/Ripple Counter
4021	8-Stage Static Shift Register
4022	Divide-by-8 Counter
4023	Triple 3-Input NOR
4024	7-Stage Binary Counter
4025	Triple 3-Input NOR
4026	Decade Counter
4027	Dual JK Flip-Flop
4028	BCD-to-Decimal Decoder
4029	Presettable U/D, B/D, Counter
4030	Quad Exclusive OR
4031	64-Stage Shift Register
4032	Triple Serial Adder (+ Logic)
4033	Decade Counter
4034	8-Stage Shift Register
4035	4-Stage Parallel-In/Parallel-Out Shift Register
4036	4-Word × 8-Bit RAM
4037	Triple AND/OR Pair

4038 Triple Serial Adder (— Logic)
4039 4-Word × 8-Bit RAM
4040 12-Stage Binary Ripple Counter
4041 Quad True/Complement Buffer
4042 Quad Clocked D Latch
4043 Quad 3-State L/S Latch NOR
4044 Quad 3-State NAND R/S Latch
4045 21-Stage Counter/Buffer
4046 Phase-Locked Loop
4047
4048 8-Input Expandable 3-State Gate
4049 Hex Inverting Buffer
4050 Hex Noninverting Buffer
4051 8-Input Analog Multiplexer
4052 Differential 4-Input Analog Multiplexer
4053 Triple 2-Channel Multiplexer
4054 4-Line LCD Driver
4055 BCD-to-7-Segment Decoder/Driver
4056 BCD-to-7-Segment Decoder/Driver
4057 4-Bit Arithmetic Array
4058
4059 4-Decade Divide by N
4060 14-Stage Counter/Oscillator
4061 256-Word × 1-Bit RAM
4062 200-Bit Shift Register
4063 4-Bit Magnitude Comparator
4064
4065 Quad Bilateral Switch
4066 Quad Bilateral Switch
4067
4068 8-Input NAND
4069 Hex Inverter
4070 Quad Exclusive OR
4071 Quad 2-Input AND
4072
4073 Triple 3-Input AND
4077 Quad Exclusive NOR
4078 8-Input NOR
4081 Quad 2-Input OR
4082 Dual 4-Input AND
4085 Dual 2-Input × 2-Input AND-OR-INVERT
4086 Expandable 4-Wide 2-Input AND-OR-INVERT
4093 Quad 2-Input NAND Schmitt Trigger
4099 8-Bit Latch
40104 TTL to High-Level CMOS Converter
40518 Dual 4-Bit Decade Counter
40527 BCD Rate Multiplier
40555 Dual 1-of-4 Decoder
40725 16 × 4-Bit RAM
40702 Bit Rate Generator
40720 256 × 1 RAM
40723 Dual 4-Bit Addressable Latch
40724
40725 16 × 4-Bit RAM

14500 SERIES

14500
14501 Triple Gate
14502 Strobed Hex Inverter/Buffer
14503 Hex Bus Driver
14504
14505 64-Bit Static RAM
14506 Dual Expandable AND-OR-INVERT
14507 Quad Exclusive OR
14508 Dual 4-bit Latch
14509
14510 BCD Up/Down Counter
14511 BCD-to-7-Segment Decoder/Driver
14512 8-Channel Data Selector
14513
14514 4/16 Line Decoder (H)
14515 4/16 Line Decoder (L)
14516 Binary Up/Down Counter
14517 Dual 64-Bit Static Shift Register
14518 Dual BCD Up Counter
14519 4-Bit AND-OR Selector
14520 Dual Binary Up Counter
14521 24-State Frequency Divider
14522 BCD Divide-by-N Counter
14523
14524 1024-Bit ROM

14525	
14526	Binary Divide-by-N Counter
14527	BCD Rate Multiplier
14528	Dual Monostable Multivibrator
14529	Dual 4-Channel Multiplexer
14530	Dual 5-Input Majority Logic Gate
14531	12-Bit Parity Tree
14532	8-Bit Priority Encoder
14533	
14534	Real Time 5-Decade Counter
14535	
14536	Programmable Timer
14537	256-Bit Static RAM
14538	
14539	Dual 4-Channel Digital Mix
14540	
14541	Oscillator-Timer
14542	
14543	BCD-to-7-Segment Decoder/Driver
14544	
14549	Successive Approximation Register
14552	64-Word × 4-Bit RAM
14553	3-Digit BCD Counter
14554	2 × 2-Bit Parallel Binary Multiplier
14555	Dual Binary 1-of-4 Decoder (H)
14556	Dual Binary 1-of-4 Decoder (L)
14557	1-to-64-Bit Shift Register
14558	BCD-to-7-Segment Decoder
14559	Successive Approximation Register
14560	NBCD Adder
14561	9s Complementer
14562	128-Bit Static Shift Register
14566	Industrial Time-Base Generator
14572	Hex Gate (2-Input NAND + 2-Input NOR + 4 Inverters)
14580	4 × 4 Multiport Register
14581	4-Bit Arithmetic Logic Unit
14582	Look-Ahead Carry Block
14583	Dual Schmitt Trigger
14584	Hex Schmitt Trigger
14585	4-Bit Magnitude Comparator

4400 AND 14400 SERIES

4402	Dual 4-Input Expandable NOR
4404	8-Stage Binary Counter
4410	2-of-8 Tone Encoder
4411	Bit Rate Frequency Generator
4412	Dual 4-Input Expandable NAND
4415	Quad Precision Timer/Driver
4416	Quad Bilateral Switch (2 NO + 2 NC)
4426	Decade Counter
4428	1-of-8 Decoder
4433	Decade Counter
4435	3½ Digit A/D Logic Subsystem
4441	Quad Low-Impedance Buffer
4445	21-Stage Counter/Buffer
4449	Hex Inverter
4490	Hex Contact-Bounce Eliminator
4440	LCD Watch Circuit
4450	Watch Oscillator Driver
4451	Watch Oscillator Driver

54C/74C CMOS

74C00	Quad 2-Input NAND
74C02	Quad 2-Input NOR
74C04	Hex Inverter
74C08	Quad 2-Input AND
74C10	Triple 3-Input NAND
74C14	Hex Schmitt Trigger
74C20	Dual 4-Input NAND
74C30	8-Input NAND
74C32	Quad 2-Input OR
74C42	BCD-to-Decimal Decoder
74C48	BCD-to-7-Segment Decoder
74C73	Dual JK Flip-Flop
74C74	Dual D Flip-Flop

74C76 Dual JK Flip-Flop with Clear and Preset
74C83 4-Bit Binary Full Adder
74C85 4-Bit Magnitude Comparator
74C86 Quad 2-Input Exclusive OR
74C89 64-Bit 3-State RAM
74C95 4-Bit Right/Left Shift Register
74C107 Dual JK Flip-Flop
74C151 8-Channel Digital Multiplexer
74C154 4-Line to 16-Line Decoder/Demultiplexer
74C157 Quad 2-Input Multiplexer
74C160 Decade Counter
74C161 Binary Counter
74C162 Decade Counter
74C163 Binary Counter
74C164 8-Bit Parallel-Out Shift Register
74C165 8-Bit Parallel-Load Shift Register
74C173 Quad D Flip-Flop, 3-State
74C174 Hex D Flip-Flop
74C175 Quad D Flip-Flop
74C192 4-Bit, Up/Down Decade Counter
74C193 4-Bit Up/Down Binary Counter
74C194 4-Bit Right/Left Shift Register
74C195 4-Bit Register
74C200 256-Bit 3-State RAM
74C221 Dual Monostable Multivibrator
74C283 4-Bit Binary Full Adder
74C901 Hex Inverting TTL Buffer
74C902 Hex Noninverting TTL Buffer
74C903 Hex Inverting PMOS Buffer
74C904 Hex Noninverting PMOS Buffer

APPENDIX **B**

CMOS Characteristics

The following tables contain much useful information about the operating characteristics of many CMOS devices. Table B-1 covers CMOS flip-flops and latches; Table B-2, CMOS shift registers (and others); Table B-3, CMOS counters; and Table B-4, CMOS decoders, multiplexers, and demultiplexers.

Table B-1. CMOS Flip-Flops and Latches

					Type						Set or Preset			Reset or Clear			Clock							Outputs		
																			Edge Trig			Rate (MHz) (Typical)				
	Device	Dual	Quad	Hex	D	JK	JK-MS	RS	Latch	Input Disable	Ind	Com	Asyn	Ind	Com	Asyn	Ind	Com	Polarity Control	Pos	Neg	+5 V	+10 V	3-State	Q	$\overline{Q}$
54C/74C	73	•					•							•		•	•				•				•	•
	76	•					•				•		•	•		•	•				•				•	•
	107	•				•								•		•	•				•				•	•
	173		•		•				•	•								•		•		4	12	•	•	
	174			•	•															•		6.5	12		•	
	175		•		•													•		•					•	•
4000	4013	•			•						•		•	•		•	•			•		4	10		•	•
	4027	•					•				•		•	•		•	•			•		3	8		•	•
	4042		•		•				•									•	•	•	•	6	10		•	•
	4043		•						•		•		•	•		•						6	10	•	•	
	4044		•						•		•		•			•						6	10	•	•	
	4076		•		•				•								•				•			•		
	14508	•							4																	

Table B-2. CMOS Shift Registers (and Other Registers)

					Inputs															Outputs				
					Shift Data						Clear				Clock					Data				
															Edge Trig			Max Freq (Typ) MHz						
	Device	Dual	Quad	Bits	Left	Right	Serial	Parallel	Shift/Load	$\overline{JK}$	Asyn	Syn	Mode Control	Par/Set Control	Pos	Neg	Inhibit	+5 V	+10 V	Ser	Par	True/Comp	3-State	Notes
54C/74C	95			4	•	•	•	•					•			2		4.5	10	•	•			
	164			8		•	2				•				•			3	8		8			
	165			8		•	•	8	•						•		•	6	12	•				
	195			4		•	2	4	•	•	•							3.0	8.5	•	5			
4000	06		•	4/5		•	•									•		3	6	•				2 Units 4 Bits 2 Units 5 Bits
	14			8		•	•	8						•	•			2.5	5	•	3			
	15	•		4		•	•				•				•			2.6	5	•	4			
	21			8		•	•	8						•	•			2.5	5	•	3			
	34			8		•	1	8					3	•				2.5	5		8			
	35			4		•		4		•	•										4	•		T/C Input

14500	17	• [1]		•	4						1		•			3	6.7	4		•	
	49	8			1				•		1		•			1.5	3.0		10		[2]
	57	1 to 64		•	2				•		7		•	•		2.5	8.0	•		•	
	59	8			1						1		•			1.5	3.0		10		[2]
	62	128		•	1								•			1.9	5.0		8		
	80	16				4							•						8	•	Multiport Register

[1] : 16, 32, 48, or 64

[2] Successive Approximation Type

Table B-3. CMOS Counters

											Count		Trig Edge					Clear or Reset		Outputs						Max Clock Freq, Typical MHz					
		Stages	Binary	BCD	Decade	Up	Down	Ripple Carry	Divide by N	Presettable	Asyn	Syn	Pos	Neg	Carry Input	Ck Enable Input	Display Enable Input	Asyn	Syn	Decoded	Carry Out	Borrow Out	Display Enable	Bipolar	Other	5 V	10 V	Notes	Lamp Test	RBI	RBO
54C/74C	160	4			•	•				•		•	•			•		•			•				4	3	8.5				
	161	4	•			•				•		•	•			•		•			•				4	3	8.5				
	162	4			•	•				•		•	•			•			•		•				4	3	8.5				
	163	4	•			•				•		•	•			•			•		•				4	3	8.5				
	192	4			•	•	•			•			•					•			•	•			4	4	10				
	193	4	•			•	•			•			•					•			•	•			4	4	10				
4000	17	5			•							•				•				10	1					2.5	5	Johnson			
	18	5							10	•		•	•													2.5	5	Johnson			
	20	14	•			•		•						•											14	2.5	7				
	22	4							8			•				•				8	1					2.5	5	Johnson			
	24	7	•			•								•											7	2.5	7				
	26	5			•											•	•	•		7	1		•		2	2.5	5	Johnson + 7-Seg Dec			
	29	4	•		•	•	•			•		•	•			•				4	1					2.5	5				
	33	5			•								•			•				7					1	2.5	5	Johnson + 7-Seg Dec	•	•	•

14400 & 4400	04	8	•			•								•				•							8	3	6.5		
	26	5			•								•			•	•	•		7	1		•	•	2	3	5	Johnson + 7-Seg Dec	• • •
	33	5			•								•			•	•			7				•	1	3	5	Johnson + 7-Seg Dec	
14500 & 4500	10	4		•		•	•			•	•		•		•						1				4	2.5	5		
	16	4	•			•	•			•	•		•		•						1				4	2.5	5		
	18	4		•							•		•	•												3	6	Dual	
	20	4	•								•		•	•							1				4	3	6	Dual	
	21	24	•											•				•							9	3.5	9		
	22	4		•					•	•			•	•		•		•							5	2.0	5.0		
	26	4	•						•	•			•	•		•		•							5	2.0	5.0		
	53			•										•						4					3	1.5	5.0	3-Digit	

Table B-4. CMOS Decoders, Multiplexers, and Demultiplexers

		Device	Decoder	Multiplexer	Demultiplexer (Data Selector)	Analog Switch	Data Inputs	RBO/BI	RBI	Lamp Test	Strobe	Code or Address Inputs	Outputs: Decoded	Outputs: Selected Output H or L	Outputs: Y	Outputs: W ($=\overline{Y}$)	Outputs: 3-State	t_{PD}[1] (approx) ns: H to L	t_{PD}[1] (approx) ns: L to H	Notes
54C/74C	42	BCD to Decimal	•									4	10	L				400	400	
54C/74C	48	BCD to 7-Segment	•					•	•	•		4	7	H				1500	1500	
54C/74C	151	8-Channel		•			8				•	3			•	•		300	300	
54C/74C	154	4-Line to 16-Line	•		•						2	4	16	L				400	400	
54C/74C	157	Quad 2-Input		•			8					2	4					250	250	
4000	16	Quad				•														10-MHz Switch Rate
4000	28	BCD to Dec Bin to Octal	•									4/3	10/8	H				700	290	
4000	54	4-Line LCD Dr	•				4				4				4					Dis F Input
4000	55	BCD to 7-Segment	•									4	7							Dis F Input Dis F Output
4000	56	BCD to 7-Segment	•								1	4	7							Dis F Input

4400	28	1 of 8	•									3	8	H				480	180	
	16	Quad				•														(2NO, 2NC) 10-MHz Switch Rate
14500	11	BCD to 7-Segment	•							•		4	7				•	720	640	BI and Latch Enable Inputs
	12	8-Channel		•			8					3					•	225	225	Disable and Inhibit Inputs
	14	4-Line to 16-Line (H)	•				4				•		16	H				800	800	Latch Inhibit Input
	15	4-Line to 16-Line (L)	•				4				•		16	L				800	800	Latch Inhibit Input
	29	Dual 4-Channel		•			8				2	2			2		•	20	20	
	43	BCD to 7-Segment	•							•		4	7					475	550	Ph (Phase) BI and Latch Disable Inputs
	55	Dual Binary 1 of 4	•	•								2	4	H				200	200	Enable Input
	56	Dual Binary 1 of 4	•	•								2	4	L				225	225	Enable Input
	58	BCD to 7-Segment	•						•	•		4	7	H				750	525	Enable Input and RBO

[1] : Propagation delays (t_{PD}) are "typical" for 5-V operation. Minimum at 5 V may be 1/3. Typical at 10-V operation is also about 1/3. Minimum at 10 V may be 1/3 of typical or $1/3 \times 1/3 = 1/9$ that shown. See device data sheets in critical cases.

Index